# Green Energy Advances

*Edited by Diana Enescu*

Published in London, United Kingdom

IntechOpen

*Supporting open minds since 2005*

Green Energy Advances
http://dx.doi.org/10.5772/intechopen.77501
Edited by Diana Enescu

Contributors
Carlos Raymundo, Juan Arevalo, Grimaldo Quispe, Fernando Almeida Prado Jr., Edvaldo Avila, Yuehong Lu, Zhijia Huang, Radu Porumb, George Serițan, Diana Enescu, Filippo Spertino, Alessandro Ciocia, Paolo Di Leo, Angela Russo, Gabriele Malgaroli, Antonio Dangola, Angelo Masi, Vincenzo Manfredi, Marianna Mecca



First published in London, United Kingdom, 2019 by IntechOpen
IntechOpen is the global imprint of INTECHOPEN LIMITED, registered in England and Wales, registration number: 11086078, The Shard, 25th floor, 32 London Bridge Street
London, SE19SG – United Kingdom
Printed in Croatia

British Library Cataloguing-in-Publication Data
A catalogue record for this book is available from the British Library

Additional hard copies can be obtained from orders@intechopen.com

Green Energy Advances
Edited by Diana Enescu
p. cm.
Print ISBN 978-1-78984-199-2
Online ISBN 978-1-78984-200-5

For EU product safety concerns:
IN TECH d.o.o., Prolaz Marije Krucifikse Kozulić 3, 51000 Rijeka, Croatia, info@intechopen.com or visit our website at intechopen.com.

# Meet the editor

Diana Enescu graduated in Industrial Energy Engineering at the University Politehnica of Bucharest (UPB), Romania, and holds a Master Degree in energy efficiency and economy of energy from UPB. She obtained her PhD in mechanical engineering at the Technical University of Civil Engineering of Bucharest. In 2006 she was a visiting postdoctoral researcher at Yale University, USA. She has published more than 40 papers in national and international journals and conferences, four books and three book chapters. She is a member of ASME (American Society of Mechanical Engineers) and a reviewer of many international journals. She teaches thermal engineering courses at the Valahia University of Targoviste, Romania. Her research interests include heat transfer, thermodynamics, and thermoelectric system applications.

# Contents

# Preface

The idea to prepare this book came from the intention to collect some of the experiences and solutions that are emerging in one of the timely areas of development of the energy systems – green energy. Nowadays there is more attention on the use of technologies and solutions that have low carbon emissions and reduce the presence of hazardous pollutants in the global and local environments. Green energy is a term widely used to group together all these technologies and solutions. A general definition of this term is a sustainable energy source that has zero or minimal environmental and economic impact, and can be obtained from solar, wind, hydro, geothermal, biomass, and other renewable energy sources. Green energy is considered as an important matter for technological, social, and industrial development, as well as to the global energy security. Green energy minimizes the negative effects of global emissions from electricity production and fossil energy resources, reduces the greenhouse effect, thereby offering a great opportunity for satisfying clean energy requirements in residential, industrial, agricultural, and commercial applications.

Significant impacts occur at the local system level, as well as for energy integration in network-based solutions. Remarkable aspects include the evolution of the buildings in to more energy efficient solutions, considering individual or grouped zero-energy buildings supplied by green energy. The increase of green energy use in the future has a positive influence on the economic increase and sustainable evolution of the society. In this case, a key element for the link between society and nature is the sustainability of green energy development and supply. Another important element to raise the life standard of people in a country is the use of low-cost green energy to support technological growth and industrial production. For this reason, researchers all over the world are involved in analysing different global stability scenarios and green energy strategies to promote and intensify the use of green energy sources and their impacts on energy security.

The essential features of the green energy sources include their compatibility with the environment, which makes these sources attractive for widespread utilisation in the context of sustainable development, while also avoiding the growth in carbon emissions and deforestation. A further main reason for wider adoption of green energy sources and technologies with great influence on sustainable development is the flexibility of their utilisation, which can be achieved in combination with appropriate storage solutions in order to mitigate the effects of the green energy source uncertainty. From the social point of view, the utilisation of local resources offers advantageous solutions and benefits to small remote areas that are not connected to the grid. In this way, the traditional paradigm of electrification coming from wide network structures can be revisited by providing electrification from local energy systems supplied with green energy, which can evolve to grid-connected systems through their integration into micro-grids.

Most of the concepts indicated above are addressed in the chapters of this book. The contents are presented by conceptually proceeding from small solutions to large energy systems:

- Green energy harvesting indicates how to extract even small amounts of energy from different physical phenomena and convert them into electrical energy. Chapter 1 addresses the specific case of thermoelectric energy harvesting, which exploits temperature differences through the Seebeck effect to provide electricity in small-scale applications.

- At the building level, considerable attention is given to reach the situation in which the building design and the contribution from local renewable energy sources will enable the construction of zero-energy buildings. Chapter 2 illustrates various aspects of the development of zero-energy buildings, starting from the definitions and addressing the design criteria for stand-alone or grid-connected solutions, the effects of possible incentives, and the uncertainty on the local renewable energy production.

- The increase in the energy efficiency of the buildings has to be considered together with other objectives, in particular when the buildings are constructed in critical environments such as in seismic areas. Chapter 3 presents an integrated approach to improve the seismic performance and to reduce energy losses of the buildings at the same time.

- The mitigation of the effects of uncertainty in renewable energy generation can also be achieved by exploiting storage solutions. Chapter 4 deals with the development of a smart battery management system used in a grid-connected photovoltaic system with storage for residential users. The selection of the best strategy to discharge the batteries, based on the electricity consumption and production forecasts, aims at reducing the peaks of power absorption from the grid.

- The evolution of renewable energy sources and storage technologies is leading the transformation of the consumers into prosumers with potential capabilities to manage their local energy systems. Chapter 5 overviews energy management aspects and standards, and discusses the evolution of local electrical energy systems with their complex interactions addressed through the Internet to ensure energy efficiency, low emissions, power quality, and secure energy trading.

- In the agricultural context, energy sustainability can be improved by the adoption of clean biomass technology. Chapter 6 addresses the usage of biomasses from agricultural wastes to reduce deforestation. The recycling of waste products from the rice industry is discussed under the prospect of contributing to the creation of a circular economy with the production and commercialization of rice husk briquettes.

- On the global system side, different climate change policies and incentives have been applied to reduce greenhouse gas emissions. Chapter 7 reports on the experience of the largest Brazilian power trader in the development of voluntary certifications associated with energy trading, which take into account the electrical energy provided by using only renewable energy sources.

This book contributes to understanding the development and application of green energy solutions. The editor wishes to thank InTechOpen for the opportunity given to prepare the book, and all the authors for the insightful contents included in their book chapters.

**PhD Diana Enescu**
Valahia University of Targoviste,
Targoviste, Romania

Chapter 1

# Thermoelectric Energy Harvesting: Basic Principles and Applications

*Diana Enescu*


## Abstract

Green energy harvesting aims to supply electricity to electric or electronic systems from one or different energy sources present in the environment without grid connection or utilisation of batteries. These energy sources are solar (photovoltaic), movements (kinetic), radio-frequencies and thermal energy (thermoelectricity). The thermoelectric energy harvesting technology exploits the Seebeck effect. This effect describes the conversion of temperature gradient into electric power at the junctions of the thermoelectric elements of a thermoelectric generator (TEG) device. This device is a robust and highly reliable energy converter, which aims to generate electricity in applications in which the heat would be otherwise dissipated. The significant request for thermoelectric energy harvesting is justified by developing new thermoelectric materials and the design of new TEG devices. Moreover, the thermoelectric energy harvesting devices are used for waste heat harvesting in microscale applications. Potential TEG applications as energy harvesting modules are used in medical devices, sensors, buildings and consumer electronics. This chapter presents an overview of the fundamental principles of thermoelectric energy harvesting and their low-power applications.


**Keywords:** thermal energy, Seebeck effect, thermoelectric generator, thermoelectric materials, design, low-power applications

## 1. Background about energy harvesting

Energy harvesting represents the energy derived from ambient sources that is extracted and directly converted into electrical energy. This way to provide energy is further used when another energy source is not available (off-grid use) to supply small- and medium-sized electronic devices, as well as electrical systems, with power from nW to hundreds of mW [1, 2]. Generally, energy harvesting refers to an environment with regular and well-assessed ambient energy sources. Energy harvesting is applied when there is a match between the available energy and the energy required.

Another term, *energy scavenging*, refers to an environment with strong non-uniform and unknown ambient energy sources [3]. Some examples of differences between the two terms are presented in **Table 1**.

The ambient energy sources used for energy harvesting are temperature gradient, electromagnetic radiation, light, motion and chemical energy (**Figure 1**).

| | Scavenging | Harvesting |
|---|---|---|
| Thermal | Forest fires | Furnace covers |
| Photonic | Interior lighting | Diurnal solar cycles |
| Mechanical | Foot traffic | Motors, ductwork |

**Table 1.**
*The difference between the terms "scavenging" and "harvesting" [3].*

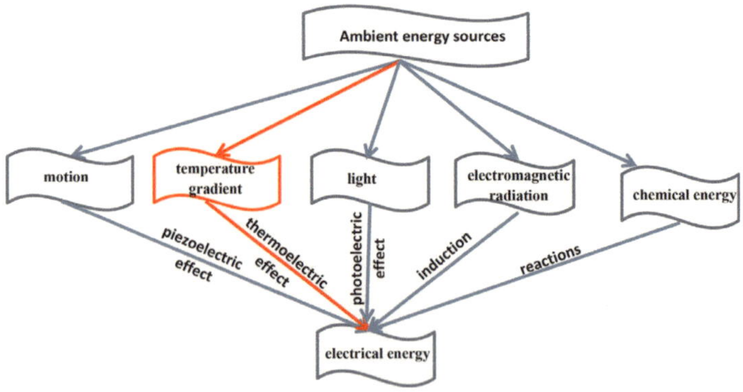


**Figure 1.**
*Energy harvesting sources.*

An energy harvester consists of:

- an energy source (which is natural or artificial);
- one or more transducers that convert environmental energy into electrical energy;
- an energy storage device (e.g., a rechargeable battery or a capacitor that stores the harvested energy);
- process control electronics [4].

The most used energy harvesters are: thermal harvester based on the thermoelectric effect; light harvester based on the photoelectric effect; electromagnetic harvester based on induction; chemical harvester based on different reactions on the electrodes surfaces; piezoelectric harvester based on mechanical vibrations or human motion (which converts pressure or stress into electricity); radio-frequency (RF) harvester (that captures the ambient radio-frequency radiation).

Thermoelectric energy harvesting mainly depends on the operation of the thermoelectric generator (TEG). A TEG converts heat directly into electrical energy according to the Seebeck effect. In this case, the motion of charge carriers (electrons and holes) leads to a temperature difference across this device. Its operation is described in Section 2.3. Furthermore, the thermoelectric energy harvesting system can generate power from hundreds of μW to mW for different sensors and transmitters.

In the last decades, the specialists' attention has been focused on the development of green energy technology to decrease fossil fuel utilisation and greenhouse gas emissions. A thermoelectric harvester produces green energy for energy

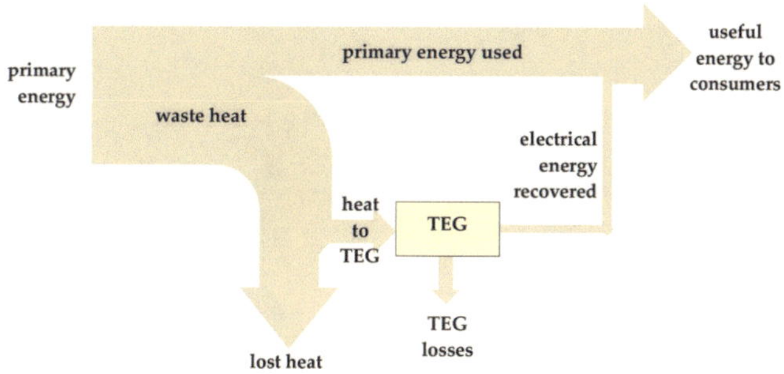


**Figure 2.**
*Electrical energy recovered from waste heat.*

harvesting with a multitude of advantages: maintenance-free, because of the use of highly reliable and compact solid-state device; silent and quiet; highly efficient in environmental terms because the heat is harvested from waste heat sources and converted into electricity; operation with high maximum temperatures (up to 250°C); useful scalable applications configured to harvest wide amounts of energy when necessary; possibility to harvest power from both hot surface or cold surface; green energy behaviour, being eco-friendly [5]. A TEG device produces energy without using fossil fuel, leading to a reduction of greenhouse gas emissions.

Unlike thermodynamic PV systems or conventional heat engines (Rankine, Stirling), the energy conversion efficiency of the TEG is limited to about 5–15% [6]. The temperature difference across the TEG system and the dimensionless thermoelectric figure-of-merit ($ZT$) have a major impact on the energy conversion efficiency [7]. It is desirable to obtain the maximum electric output power and efficiency when a TEG system operates. In case of waste heat recovery applications [8], only electric output power is significant and the heat not recovered is lost [9]. Considering that thermal energy harvesting has a reduced efficiency (5–6%), this could represent a major barrier for its extensive utilisation. An improvement in the TEG efficiency bigger than 10% has been lately obtained due to the progress of new thermoelectric materials [10].

The recovery of the electrical energy from waste heat using diverse sources is depicted in **Figure 2**.

## 2. Basic principles of thermoelectric energy generation

### 2.1 Thermoelectric effects

The thermoelectric effects are reversible phenomena leading to direct conversion between thermal and electrical energy [9]. Direct energy conversion relies on the physical transport properties of the thermoelectric materials (thermal conductivity, electric conductivity and Seebeck coefficient) and their energy conversion efficiency in terms of the figure-of-merit. These materials are suitable to convert thermal energy into electrical energy and vice-versa. The main phenomena that occur in a thermoelectric device are the thermoelectric effects (Seebeck, Peltier, Thomson), and the Joule effect.

- When the electrical energy is converted into thermal energy, the phenomenon is known as the *Peltier effect*, with applications in cooling and heating. The device used in such applications is called thermoelectric cooler (TEC) [11–13]. In this case, thermoelectric modules are efficient temperature controllers [14].

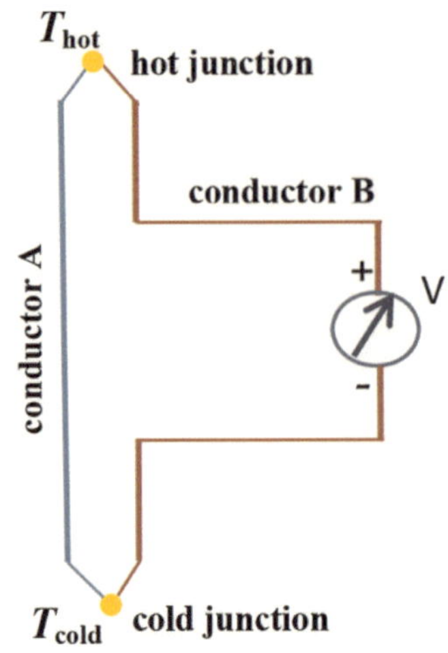


**Figure 3.**
*Schematic of the Seebeck effect in an open circuit.*

- When the thermal energy is converted into electrical energy, the phenomenon is known as the *Seebeck effect*, with applications for power generation. The device used in such applications is called thermoelectric generator (TEG) [15, 16].

The *Seebeck effect* occurs when a temperature difference across a conductor provides a voltage at the conductor ends. Two distinct conductors A and B are linked together to compose the junctions of a circuit (**Figure 3**). These conductors are connected electrically in series and thermally in parallel. One junction has the hot temperature $T_h$ and another one has the cold temperature $T_c$, with $T_h$ bigger than $T_c$. The Seebeck effect appears due to the thermal diffusion which provokes the motion of the charge carriers (electrons or holes) across (or against) temperature difference in the conductors.

The Seebeck voltage at the circuit junctions can be written as:

$$V = \left( \underbrace{\alpha_A - \alpha_B}_{\alpha_{AB}} \right) \cdot \left( \underbrace{T_h - T_c}_{\Delta T} \right) \quad (1)$$

where $\alpha_A$ and $\alpha_B$ are the Seebeck coefficients for the conductors A and B, in $V \cdot K^{-1}$.

The Seebeck coefficient of a thermoelectric material or *thermopower* $\alpha_{AB}$ is the connection parameter between the input temperature difference and the output voltage difference. The Seebeck coefficient of a thermoelectric material depends on temperature, as well as on other two physical transport properties (thermal conductivity, electric conductivity). It determines the thermoelectric material performance. Its magnitude ranges from $\mu V \cdot K^{-1}$ to $mV \cdot K^{-1}$ and depends on the junction temperature, and its sign is influenced by the semiconductor material [17]. Furthermore, the sign of the Seebeck coefficient depends on the type of carriers (electrons $e^-$ and holes $h^+$) conducting the electric current. If the electric current is conducted by $e^-$, the sign of the Seebeck coefficient is negative. If the electric current is conducted by $h^+$, the sign of the Seebeck coefficient is positive [18].

The Seebeck coefficient $\alpha_{AB}$, the temperature gradient $\nabla T$, and the electric field $E$ are written under the following relationship:

$$\alpha_{AB} = \frac{E}{\nabla T} \tag{2}$$

The *Thomson effect* affirms that in any conductive material in which the electrical current flows in the presence of a temperature difference between two ends, heat is also released or absorbed. The Thomson heat released or absorbed is given as:

$$\dot{Q} = \underbrace{\rho \cdot J^2}_{\text{Joule heating}} - \underbrace{\mu_{AB} \cdot J \cdot \nabla T}_{\text{Thomson heating}} \tag{3}$$

where $\rho = \frac{1}{\sigma}$ is the electrical resistivity in $[\Omega \cdot \mathrm{m}]$, $\sigma$ is the electrical conductivity in $[\mathrm{S} \cdot \mathrm{m}^{-1}]$, $J$ is the current density in $[\mathrm{A} \cdot \mathrm{m}^{-2}]$, $\mu_{AB}$ is the Thomson coefficient in $[\mathrm{V{\cdot}K^{-1}}]$, and $\nabla T$ is $\nabla T = \frac{\mathrm{d}T}{\mathrm{d}x}$ is the temperature gradient along the conductor in [K.]

*Joule heating* occurs when an electric current that flows through a conductor produces heat. Joule heating does not change its sign in Eq. (3), while Thomson heating (the second term) changes its sign, following $J$.

Therefore, the sign convention of the Thomson coefficient is considered as [17]:

- *positive* when the current flows from the low-temperature side to the high-temperature side of the conductor and the heat is absorbed through it;
- *negative* when the current flows inversely and the heat is rejected from it;
- *null* when the current flows from the high to the low side and vice-versa and the heat is neither generated nor absorbed.

The following relationships hold between the Seebeck coefficient and the Peltier coefficient, as well as between the Seebeck coefficient and the Thomson coefficient. These are called Thomson relations [14]:

$$\pi_{AB} = \alpha_{AB} \cdot T \tag{4}$$

$$\mu_{AB} = T \cdot \frac{\mathrm{d}\alpha_{AB}}{\mathrm{d}T} \tag{5}$$

### 2.2 Thermoelectric effects and thermodynamic processes

Thermoelectric effects that take place in TEG devices are subject to the thermodynamic laws. According to thermodynamics, the heat transfer across a finite temperature difference is an irreversible process and the entropy change of such process is positive. The heat conduction and Joule heating are considered as irreversible processes.

The heat is irreversibly produced according to the Joule effect when an electrical current flows through a conductor or semiconductor. The Joule effect takes place at the TEG interconnects due to their electrical contact resistance or in a thermocouple. Other irreversibilities are found in the heat transfer between the TEG and the local environment [9]. If the irreversible processes are removed, the entropy becomes null. In this case, the ideal conditions given by the Carnot efficiency or *COP* (coefficient of performance) are achieved [19]. A deep overview of steady-state irreversible processes as heat conduction in semiconductor materials, metals and other solid-state devices is presented in [19, 20]. The Seebeck, Thomson and Peltier effects are reversible thermodynamic processes [21]. When the current flows through a conductor, both the Joule effect and the Thomson effect take place

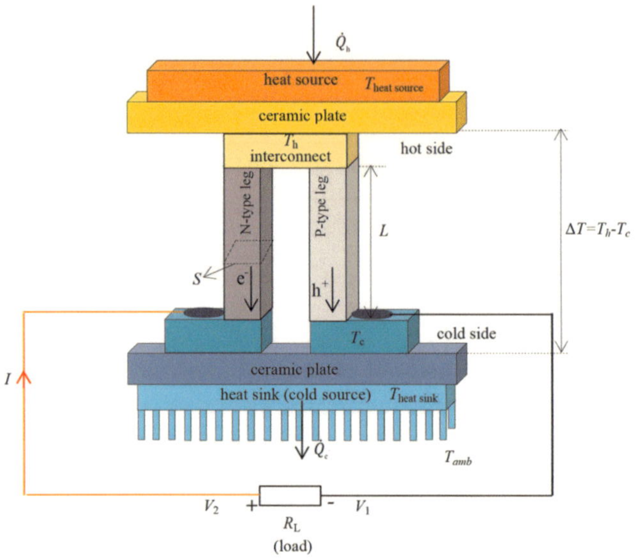


**Figure 4.**
*Schematic of a TEG device with a single thermoelectric couple and two legs.*

simultaneously, and the magnitude of the Thomson effect is about two times less than the magnitude of the Joule effect [17].

### 2.3 TEG structure and model

The TEG device is composed of one or more thermoelectric couples. The simplest TEG consists of a thermocouple, comprising a pair of P-type and N-type thermoelements or legs connected electrically in series and thermally in parallel. The differentiation between N- and P-doped materials is important. The P-type leg has a positive Seebeck coefficient and an excess of holes $h^+$. The N-type leg has a negative Seebeck coefficient and an excess of free electrons $e^-$ [22]. The two legs are linked together on one side by an electrical conductor forming a junction or interconnect, usually being a copper strip. Let us denote the voltage at the outside terminal connected to the N-type leg on the cold side of TEG as $V_2$, while the voltage at the external terminal connected to the P-type leg on the cold side of TEG is $V_1$ (**Figure 4**). An electrical load having resistance $R_L$ is connected in series with the output terminals of TEG creating an electric circuit. When the electric current flows in this electrical load, an electrical voltage is generated at its terminals. The TEG device will generate DC electricity as long as there is a temperature gradient between its sides. When the temperature difference $\Delta T = T_h - T_c$ across the TEG device increases, more electric output power will be generated.

A number of thermoelectric couples $n$ form a TEG system wired electrically in series and sandwiched between two ceramic plates to maximise the output voltage from the TEG (**Figure 5**).

In this case, the equivalent internal resistance of the thermoelectric couples in series is:

$$R = n \cdot \left[\rho_P \cdot L_P \cdot (S_P)^{-1} + \rho_N \cdot L_N \cdot (S_N)^{-1}\right] \quad (6)$$

and the equivalent thermal conductance of the thermoelectric couples in parallel is:

$$K = n \cdot \left[k_P \cdot S_P \cdot (L_P)^{-1} + k_N \cdot S_N \cdot (L_N)^{-1}\right] \quad (7)$$

where $\rho = R \cdot \frac{S}{L}$ is the electrical resistivity of each leg, $S$ is the cross-sectional area of the each leg in $[m^2]$, $L$ is the leg length in [m], $k$ is the thermal conductivity of each leg in $\left[W \cdot (m \cdot K)^{-1}\right]$, and the thermal conductance of each leg is $K = k\frac{S}{L}$ in $\left[W \cdot K^{-1}\right]$

These relations are further simplified considering that N-type and P-type legs are the same as form ($L = L_P = L_N$ and $S = S_P = S_N$) and material properties ($\rho = \rho_P = \rho_N$, and $k = k_P = k_N$). The equivalent internal resistance becomes:

$$R = n \cdot 2\rho \cdot L \cdot (S)^{-1} \quad (8)$$

and the equivalent thermal conductance is:

$$K = n \cdot 2\,k \cdot S \cdot (L)^{-1} \quad (9)$$

If the electrical contact resistance $R_a$ is not negligible, the equivalent internal resistance of the thermoelectric couples in series becomes:

$$R = n \cdot 2\rho \cdot L \cdot (S)^{-1} + R_a \quad (10)$$

The voltage at the TEG terminals is:

$$V_{TEG} = V_2 - V_1 = n \cdot (I \cdot R - \alpha_{PN} \cdot \Delta T) = n \cdot I \cdot R - V_{Seebeck} \quad (11)$$

where $\alpha_{PN} = (\alpha_P - \alpha_N)^2$ is the Seebeck coefficient of the thermoelectric couple. The input electrical current in the circuit is:

$$I = \frac{V_{Seebeck}}{n \cdot R + R_L} = \frac{n \cdot \alpha_{PN} \cdot \Delta T}{n \cdot R + R_L} \quad (12)$$

where the load resistance $R_L$ is connected to the output of the circuit where the electric output power generated by TEG is consumed; the Seebeck voltage is $V_{Seebeck} = V_P - V_N = \alpha_{PN} \cdot \Delta T$. The relationship between $V_{Seebeck}$ and $\Delta T$ is non-linear, therefore $\alpha_{PN}$ depends on temperature.

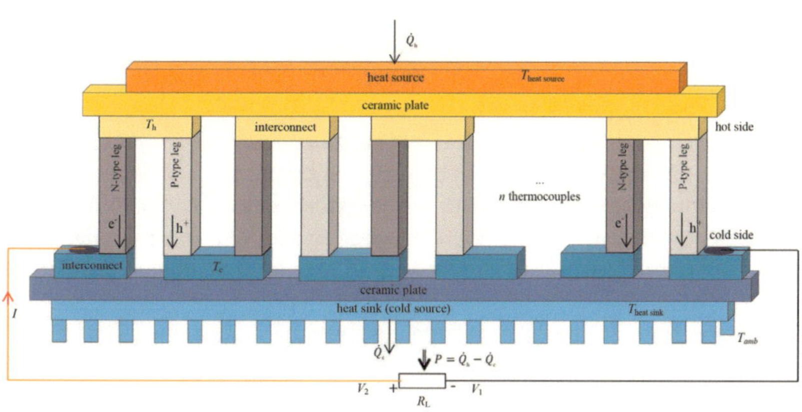


**Figure 5.**
*Schematic of a TEG device with* n *thermoelectric couples.*

The electric output power delivered by TEG to the load is:

$$P = n \cdot \left(\alpha_{PN} \cdot I \cdot \Delta T - R \cdot I^2\right) \tag{13}$$

On the other side, the electric output power absorbed by the load (considering the conventional sign, with the current flowing as indicated in **Figure 5**) is:

$$P = -V_{TEG} \cdot I = n \cdot \left(R \cdot I^2 - \alpha_{PN} \cdot I \cdot \Delta T\right) \tag{14}$$

The electric output power absorbed by the load resistance $R_L$ is:

$$P_R = I^2 \cdot R_L = \left(\frac{n \cdot \alpha_{PN} \cdot \Delta T}{n \cdot R + R_L}\right)^2 \cdot R_L \tag{15}$$

The maximum electric output power of a TEG is obtained when the electrical output power is maximised with respect to the electric current:

$$P_{max} = n \cdot \frac{(\alpha_{PN} \cdot \Delta T)^2}{4R} \tag{16}$$

$$I_{max} = \frac{\alpha_{PN} \cdot \Delta T}{2R} \tag{17}$$

The maximum electrical output power delivered by TEG is obtained if the load resistance is equal to the equivalent internal resistance of the thermoelectric couples in series $(R_L = R)$ [23].

The heat flow rate absorbed at the hot junction of the TEG depends on the Peltier heat, the heat conduction and the Joule heat. The heat flow rate absorbed at the hot junction depends on the thermoelectric material properties and leg geometries:

$$\dot{Q}_h = n \cdot \left[\alpha_{PN} \cdot T_h \cdot I - \frac{R \cdot I^2}{2} + K \cdot \Delta T\right] \tag{18}$$

A TEG could be considered as a thermal battery, a physical structure used to store and release thermal energy. The electromotive force of this thermal battery is the Seebeck voltage (**Figure 6**).

### 2.4 Components of a thermoelectric energy harvesting system

A thermoelectric energy harvesting system consists of the following parts (**Figure 7**):

- *Thermoelectric generator* (TEG): if $\Delta T$ is kept between the hot and cold sides of the device, an external circuit can be supplied by the voltage resulting at the TEG output terminals, providing power to the external electrical load. A single TEG generates power from 1 to 125 W. The use of more TEGs in a modular connection may increase the power up to 5 kW and $\Delta T_{max}$ could be bigger than 70°C.

- *Heat source*, for example, a heat pipe system (the TEG devices and the heat pipe system can be used together in waste heat recovery systems). The heat pipe is a passive (no moving parts or fan) metallic device which has a high heat transfer capacity (very high thermal conductivity), with minimal thermal resistance and almost no heat loss; it operates in a medium temperature to high-temperature range; the common working fluid is water operating at a

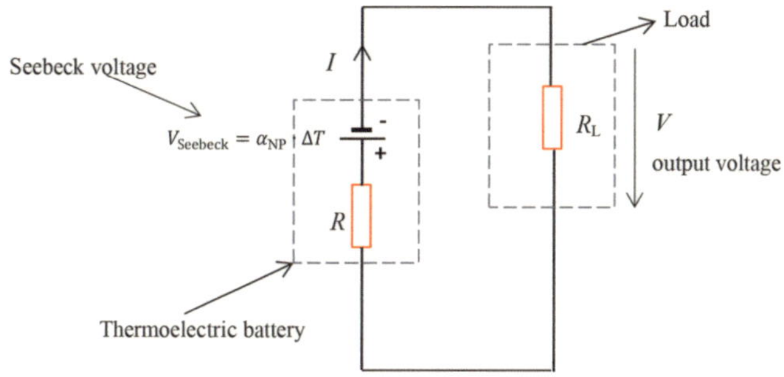


**Figure 6.**
*Equivalent circuit of a TEG device.*

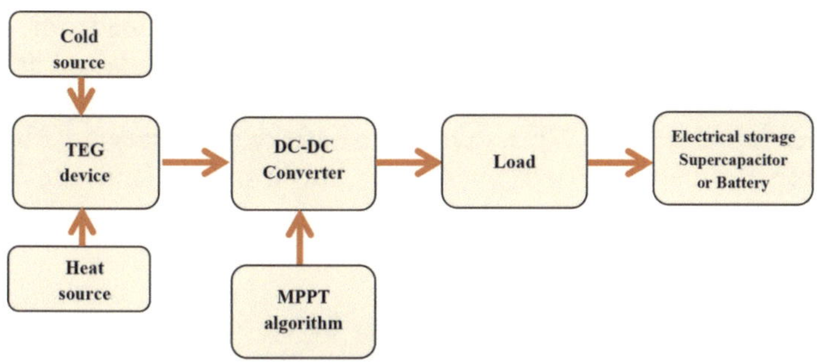


**Figure 7.**
*Block diagram of a thermoelectric energy harvesting system.*

temperature of about 300°C; for higher operating temperature ranges other working fluids are used (e.g., naphthalene or liquid metals like potassium and sodium) [24]; heat pipes are used for temperature regulation of the TEGs; in some applications (e.g., industrial glass processes) a heat exchanger can be attached on the hot side; its role is to absorb the thermal energy (e.g., from the glass process exhaust stream) and to transfer it to the TEG, which converts it partially into electrical energy; the remaining unconverted thermal energy is transferred from the TEG cold side to the cold source, and is dissipated to the environment at ambient temperature $T_{amb}$.

- *Cold source* is the heat transfer system containing heat exchangers (heat sinks, coils, cooling blocks and radiators) to enhance the heat dissipation across the TEG; this process is useful to obtain a bigger temperature difference across the TEG [7, 25]; the heat sink is a device that has the role to transfer heat from a hot surface to a fluid (gas, ambient air or liquid); the assessment and design of different heat sink types for TEG system is presented in [26]. The metal heat sink contains many fins. To increase its dissipation rate, the fins area, the heat transfer coefficient, and the fin thermal conductivity are raised.

The heat sink is required at the TEG when a high heat flow rate is applied on the TEG hot side, and the cold side is kept at low temperature, leading to high conversion efficiency; in this case, the TEG efficiency is strongly influenced by the TEG design.

- DC-DC *converter* (Boost, Buck-Boost, Buck, Sepic, or Cuk converter), which is a power electronic circuit designed for voltage conversion (to convert a DC source from one voltage level to another voltage level) [27]; since the output voltage of the TEG is low or is not constant, it is necessary to provide a DC-DC converter; its role is to increase the output voltage obtained in the TEG (which depends on the number of TEGs in series and on the TEG features) corresponding the requirements of the external load. For these DC-DC converters, accurate control is necessary. In this case, the implementation of the Maximum Power Point Tracking (MPPT) algorithm within the DC-DC converter controller is essential. To enhance the real system feasibility, it is necessary to harvest from TEGs as much electric output power as possible; the effectiveness of TEG operation could be checked by assessing the DC-DC converter operation and the MPPT control.

- *DC load*, used to be connected to a supercapacitor or to recharge a battery to store energy; the battery stores DC voltages at a charging mode and powers DC electrical energy in a discharging mode; typical DC loads for TEG like batteries operate at 12 V; the output voltage of the TEG device at the MPP (Maximum Power Point) must be higher than 12 V for example in buck converter applications [27]; to avoid the battery overcharging a battery regulator is sometimes used; the electric power output from the DC-DC converter can be stored over time in a supercapacitor, to be released to the load when needed [28].

The efficiency of the thermoelectric energy harvesting system is defined as the ratio of the electrical energy output (used or stored) to the total energy input. This efficiency also contains the electrical efficiency of TEGs, the heat exchangers efficiency, as well as the efficiency of the DC-DC converter. The total energy input especially depends on the energy obtained from the hot source. Also, the total energy input depends to a lesser extent on the mechanical energy needed to operate the thermoelectric energy harvesting system (e.g., pressure losses in the heat exchangers or cooling of the cold heat sink) [29].

Researchers are focused on the improvement of the thermoelectric conversion efficiency of TEGs. For this reason, two objectives must be fulfilled. The first objective is to improve the dimensionless figure-of-merit *ZT* by the optimisation of thermoelectric materials. The second objective is to decrease the thermal resistance between the heat source and the hot side of the TEG, as well as between the cold side of the TEG and the environment [30].

### 2.5 Efficiency assessment of a TEG device

The electrical efficiency of a TEG (or thermoelectric conversion efficiency) is the ratio between the electric output power $P$ delivered to the load and the rate of heat input $\dot{Q}_h$ absorbed at the hot junction of the TEG and transferred through the TEG. This means that a TEG converts the rate of heat input $\dot{Q}_h$ into electric output power $P$ with electrical efficiency $\eta_{\mathrm{TEG}}$ [5].

$$\eta_{\mathrm{TEG}} = \frac{P}{\dot{Q}_h} \tag{19}$$

Eq. (19) is written in more details as

$$\eta_{\mathrm{TEG}} = \frac{n \cdot R_{\mathrm{L}} \cdot \Delta T \cdot \alpha_{\mathrm{PN}}^2}{K \cdot (n \cdot R + R_{\mathrm{L}})^2 + n \cdot (R_{\mathrm{L}} \cdot T_{\mathrm{h}} + n \cdot R \cdot T) \cdot \alpha_{\mathrm{PN}}^2} \tag{20}$$

where $T$ is the absolute temperature representing the mean temperature between the cold side and hot side of the TEG and is written as $T = \frac{T_h + T_c}{2}$.

The efficiency corresponding to $P_{max}$ is:

$$\eta_{TEG} = \frac{\Delta T}{4 \cdot Z^{-1} + T_h + T} \tag{21}$$

where $Z = \frac{\alpha_{PN}^2}{K \cdot R}$ is the figure-of-merit for a thermocouple.

The thermoelectric conversion efficiency is maximised with respect to $R_L$ [23] when

$$m = \frac{R_L}{R} = \sqrt{1 + ZT} \tag{22}$$

The TEG device operates as all thermal engines with efficiency less than the efficiency of ideal Carnot cycle $\eta_C = \frac{T_h - T_c}{T_h} = \frac{\Delta T}{T_h} < 1$ [31]:

$$\eta_{TEGmax} < \eta_C \tag{23}$$

In this case, the thermoelectric conversion efficiency is limited by the Carnot efficiency and is written, by introducing the reduced efficiency $\eta_r$, as:

$$\eta_{TEGmax} = \underbrace{\frac{\sqrt{1 + ZT} - 1}{\sqrt{1 + ZT} + \frac{T_c}{T_h}}}_{\eta_r} \cdot \underbrace{\frac{\Delta T}{T_h}}_{\eta_C} = \frac{m - 1}{m + \frac{T_c}{T_h}} \cdot \frac{\Delta T}{T_h} = \frac{\eta_r}{T_h} \cdot \Delta T \tag{24}$$

and the corresponding electric output power is:

$$P(\eta_{TEGmax}) = \frac{n \cdot \sqrt{1 + ZT}}{R} \cdot \left( \frac{\alpha_{PN} \cdot \Delta T}{\sqrt{1 + ZT}} \right) \tag{25}$$

For a cold side temperature of $T_c$ = 300 K and $\Delta T$ in the range of 20 K, $\eta_{TEGmax} \cong 1\%$ is obtained [32].

As observed in Eq. (24), the TEG efficiency strongly depends on the *operating temperatures* of TEG ($\Delta T$ between the junctions), the *dimensionless thermoelectric figure-of-merit ZT*, and additionally the TEG *design* (cross-sectional area, length and shape) [33].

The TEG efficiency $\eta_{TEG}$ rises almost linearly with $\Delta T$, and the ratio $\frac{\eta_r}{T_h}$ is almost constant [5]. The bigger the temperature difference, the more efficient the TEG device will be. A TEG can work at about 20% of the Carnot efficiency over a large temperature range [24]. The TEG efficiency is about 5% and its electric output power is delivered at any $\Delta T$. If materials with $ZT$ = 10 would exist, there could be TEGs with $\eta_{TEG} = 25\%$ at $\Delta T$ = 300 K [25].

The thermoelectric waste heat recovery is influenced to a bigger extent by the thermoelectric conversion efficiency $\eta_{TEG}$, and to a lesser extent by the heat exchanger design. The ratio between thermal efficiency $\eta_t$ and thermoelectric conversion efficiency represents the fraction of waste heat passed through the thermoelectric couples, given as [34]:

$$\varepsilon = \frac{\eta_t}{\eta_{TEG}} \tag{26}$$

The maximum efficiency $\eta_{TEGmax}$ depends on the temperature difference $\Delta T_{TEG}$ at which the TEG works [31]. The maximum conversion efficiency occurs when:

$$\frac{R_L}{R} = \sqrt{1 + Z\frac{T_c + T_h}{2}} \tag{27}$$

*2.5.1 The dimensionless thermoelectric figure-of-merit ZT*

The dimensionless thermoelectric figure-of-merit $ZT$ is used to characterise a thermoelectric material performance, as well as the efficiencies of various TEGs working at the same temperatures [24].

$ZT$ depends on the physical transport properties: the thermal conductivity $k$, the electrical conductivity $\sigma = \frac{1}{\rho}$, and the Seebeck coefficient $\alpha$:

$$ZT = \frac{\alpha^2 \cdot T}{\rho \cdot k} = \frac{\alpha^2 \cdot \sigma \cdot T}{k} \tag{28}$$

The upper side term $\alpha^2 \cdot \sigma$ is called *the power factor*, a parameter that assesses the performance of a thermoelectric material.

The higher is $ZT$, more performant is the thermoelectric material and the better is the TEG. In the practical applications, the maximum $ZT$ is about 2 and corresponds to a maximum conversion efficiency of about 20% [35].

A good thermoelectric material must fulfil the following requirements:

- Seebeck coefficient as high as possible to maximise energy conversion; the generated open-circuit-voltage is proportional to the Seebeck coefficient and to the temperature difference across the TEG ($V_{Seebeck} = \alpha_{PN} \cdot \Delta T$). In this case, a high Seebeck coefficient leads to a high voltage. This condition is very important for increasing the energy conversion [22].

- Electrical conductivity $\sigma$ as high as possible in order to reduce Joule heating due to the internal electrical losses [22].

- Thermal conductivity $k$ as low as possible to maintain heat at the junctions, to allow a large $\Delta T$ maintained across the TEG, and to minimise thermal losses through the thermoelectric material [19].

The effective figure-of-merit of TEG, $ZT_{TEG}$ depends on the dimensionless thermoelectric figure-of-merit, and the specific contact electrical resistivity according to the expression:

$$ZT_{TEG} = \frac{L}{(L + 2\sigma \cdot \rho_a)} \cdot ZT = \frac{L}{(L + 2\sigma \cdot \rho_a)} \cdot \frac{\alpha^2 \cdot \sigma \cdot T}{k} \tag{29}$$

where $\rho_a = R_a \cdot S_a$ is the specific contact electrical resistivity. Ideally, for an efficient TEG $\rho_a < 1\mu\Omega \cdot cm^2$ and instead, for a typical TEG, $\rho_a < 2 \cdot 10^{-4}\Omega \cdot cm^2$ [36].

Although the low efficiency is a drawback to the progress of TEGs, researchers' and manufacturers' attention is focused on the improvement of the following characteristics:

- the dimensionless thermoelectric figure-of-merit $ZT$;

- the operating range of thermoelectric materials to work with the $\Delta T$ as high as possible;
- the use of low-price materials to reduce the negative impact of low efficiency [29].

The most popular thermoelectric material is Bismuth Telluride ($Bi_2Te_3$). Its utilisation in TEGs is limited (only for industrial modules with an average value of $ZT$ from 0.5 to 0.8) because the maximum temperature at the hot side of the devices is relatively reduced [29]. In the power generation applications, the best commercially available TEGs made of $Bi_2Te_3$ have a $ZT$ of about 1 at the temperature 300 K, leading to a low thermal efficiency of the thermoelectric device (less than 4%) [24]. The thermoelectric materials must be both stable from the chemical point of view and strong from the mechanical point of view at high temperatures (e.g., for the automotive exhaust waste heat recovery, at specific working conditions, the range of the average exhausts temperature is from 500 to 600°C with values increasing up to 1000°C) [37]. To improve the thermoelectric properties of TEG, the researchers' attention is focused on the development of new thermoelectric materials. Calcium manganese and lead telluride are the thermoelectric materials used in the TEG legs, because they resist at higher temperatures. The hot side of TEG is made of materials having a high $ZT$ at higher temperatures (e.g., lead telluride). The cold side of the TEG is made of materials having high $ZT$ at reduced temperatures (e.g., $Bi_2Te_3$) [24]. At present, even though the research of the thermoelectric materials development is focused on obtaining the high $ZT$ of 2, unfortunately the efficiency of TEG is limited to $\eta_{TEG} < 10\%$ [38]. Significant progress has been made towards increasing the thermoelectric efficiency of different inorganic material classes (e.g., skutterudites [39], tellurides [40, 41], half-Heuslers [42] and silicides [43]). The researchers' attention is focused on the development of organic materials for thermoelectric energy harvesting due to their advantages (e.g., low-cost, reliability, low weight and so on). For this reason, some polymers with different doping levels (like polyaniline (PANI), polyamide (PA), and poly (3,4-ethylenedioxythiophene) or PEDOT) are assessed for future applications [44].

To obtain high efficiency, segmented TEGs use high-temperature differences to raise the Carnot efficiency $\eta_C$ [45]. When a TEG operates with a high-temperature difference, each thermoelement of the device can be divided into multiple segments of different thermoelectric materials. In this way, each material is working in a more limited temperature range where this has a good performance [46]. The segmented design of a TEG is an efficient mode to improve its performance. In this case, two or more thermoelectric materials along the direction of the leg height are used to match the optimal temperature range of the thermoelectric material. It means that a thermoelectric material with high efficiency at raised temperature is segmented with another thermoelectric material with high efficiency at reduced temperature [45]. The maximum efficiency is obtained when the relative current density $\bar{J}$ is equal to the compatibility factor $u$ of the thermoelectric material [47]:

$$\underbrace{\frac{J}{k \cdot \nabla T}}_{\bar{J}} = \underbrace{\frac{\sqrt{1 + ZT} - 1}{\alpha \cdot T}}_{u} \tag{30}$$

The compatibility factor is used for choosing the proper material [48]. El-Genk and Sabre [46] obtained a TEG energy conversion efficiency of about 12% by using a segmented thermoelectric couple. Snyder [47] observed that the segmentation of the thermoelements with SnTe or PbTe produced low extra power, while the filled Skutterudite obtained an increment in efficiency from 10.5 to 13.6%. Further studies [47, 48] reported that the segmentation was efficient only for $u \leq 2$. Ngan et al. [49] demonstrate that segmentation reduces the total efficiency by neglecting the compatibility factor of thermoelectric materials. Hung et al. [50] showed that the performance and the power production of the segmented TEG are three times bigger than a normal TEG. The analytical assessment concerning the effect of the leg geometry on the performance of the segmented TEG was performed in [51, 52]. Their conclusion is that both power and efficiency are increased when the segmented TEG is used. Vikhor and Anatychuk [53] carried out a theoretical analysis. The results showed an efficiency of the segmented TEGs bigger than 15% compared to the non-segmented TEGs. Zhang et al. [54] proposed a design method of optimization with predictive performance to obtain maximum conversion efficiency. In this case, the segmented modules consisted of $Bi_2Te_3$-based alloys and $CoSb_3$-based skutterudites, with an efficiency of 12% when working under a $\Delta t$ = 541°C. The very low losses and the good design based on the numerical evaluation showed that the conversion efficiency was up to 96.9% of the theoretical efficiency.

### *2.5.2 TEG design for energy harvesting applications*

In TEG systems, a crucial factor is the optimisation of the systems design, together with the heat source and heat sink attached to the TEG device. Industrial utilisation of TEGs needs other components (like heat exchangers and DC-DC converter) to form a powerful TEG [29].

The TEG performance is influenced not only by the low conversion efficiency, but also by the heat transfer conditions on the cold and hot sides of TEG and its geometry. The $\Delta T$ between two junctions depends on the good heat transfer between TEGs and heat sources or heat sinks. For this reason, the design and interactions between heat exchangers and TEGs are very important problems. There are two paths to solve these problems together. The first path is the optimisation of the TEGs system. The second path is the enhancement of the heat transfer at the TEG sides [55].

#### *2.5.2.1 Optimization of the TEG device*

The TEG device optimisation is correlated with the impact of the geometry device [56]. It has been demonstrated that an important rise in the electric output power from TEG is obtained by changing the leg geometry. The leg geometry is optimised by determining the leg height and the number of thermocouples, leading to maximisation of electric output power or efficiency at given operating conditions. Therefore, there is interdependence between the optimal leg geometry and the electrical load resistance $R_L$ for a TEG. Hodes [23] presented a method to compute the leg geometry (number and height) that maximises the electric output power and $\eta_{TEG}$ with negligible or finite electrical contact resistance at TEG interconnects. If a TEG has a low number of legs, the energy conversion is low, because the $R_L$ is not sufficient to obtain an adequate high voltage. Inversely, if a TEG has too many legs, the total equivalent resistance of the TEG will increase and relatively high Joule losses will occur in the TEG when the load is supplied.

There is an optimal solution also for the leg length. If the leg is long, the electric output power is limited due to the increase in the internal resistance of the leg that limits the electric current. Conversely, a short leg will behave as a good thermal conductor that reduces the temperatures between its ends; hence, even though the internal electrical resistance will be low, the electric output power will not be significant and the electric conversion efficiency will be low [9].

Lavric et al. [57] demonstrated that the electric output power is influenced by the effects above mentioned (a reduction of the leg length leads to a reduction of the electrical resistance; an increase of the leg length leads to the higher temperature difference across the TEG). If the geometric parameters of the TEG (leg length, semiconductor pair number and the base area ratio of semiconductor columns to TEG) are optimised, the electric output power and the thermoelectric conversion efficiency are considerably improved. Such an improvement is also reported in [58]. The first step was to consider the electric output power as the objective function and the inputs were the geometric parameters. The electrical output power values were about 269, 314, 338, and 893% higher than the values of the initial design. The second step was to consider the TEG conversion efficiency $\eta_{TEG}$ as the objective function. A $\eta_{TEG}$ rise is obtained for the optimal design at the same time with an important reduction of $P$. Finally, the third step was to use multi-objective optimization to improve both $P$ and $\eta_{TEG}$, simultaneously.

Two dimensionless parameters influence the maximisation of the electric output power and the conversion efficiency of TEG [35]:

- slenderness ratio, which is a geometric parameter:

$$x = \frac{S_P \cdot (L_P)^{-1}}{S_N \cdot (L_N)^{-1}} \tag{31}$$

- external load parameter:

$$y = \frac{R_L}{L_N \cdot (\sigma_N . S_N)^{-1}} \tag{32}$$

The thermal efficiency of the TEG can be improved while decreasing the slenderness ratio for large external load parameters. Yilbas and Sahin [35] obtained high conversion efficiency for the slenderness ratio $0 < x < 1$ in the case of all the external load parameters.

Zhang et al. [55] propose a design method of thermoelectric elements segmentation of TEG, considering their length as the first design parameter. The optimal length ratio, referring to the highest values of the maximum electric output power, and the thermoelectric conversion efficiency are influenced by thermoelectric materials, leg geometry and heat transfer characteristics. Zhang et al. [59] proposed two new parameters, namely, the power factor associated with the electric output power, and the efficiency factor associated with the thermoelectric conversion efficiency. These new parameters are useful for obtaining the optimum temperature range of each segment:

$$(ZJ)_P = \alpha^2 \cdot \sigma \cdot (1 + m \cdot k)^{-2} \tag{33}$$

$$(ZJ)_\eta = Z \cdot (1 + m \cdot k)^{-1} \tag{34}$$

considering that the thermoelectric materials of the TEG legs have the same physical transport properties ($\alpha_P = -\alpha_N$; $\sigma_p = \sigma_N$; $k_P = k_N$) and $m$ is a variable factor that depends on the leg cross-sectional area, and on the heat transfer coefficients on both TEG sides.

#### 2.5.2.2 Heat transfer enhancement at the hot and cold sides

The fins attached to the heat transfer surfaces are very important for enhancing the heat transfer at the hot and cold sides. One interface is between the heat sink and the TEG cold side, and the other interface is between the heat source and the TEG hot side.

An increment of the fin height and fin number results when the electric output power of the TEG rises [60]. An optimal connection between the height and the number of fins to provide the maximum net electric output power is obtained in Jang et al. [60]. The heat transfer increases when the fin number is higher and the fin height rises, due to the extension of the heat transfer area. However, when the height of the fin increases over a given value, the change in the output electrical power becomes less significant. Borcuch et al. [61] investigated the effect of hot side heat exchanger design on the operating parameters of a TEG. Furthermore, the heat sink connected to the TEG device must be thermally matched with the TEG to maximise the electric output power and voltage. In this case, the thermal interface losses are practically negligible, that means $T_{\text{heat sink}} \cong T_{\text{c}}$ and $T_{\text{heat source}} \cong T_{\text{h}}$.

To maximise the output voltages of TEG, a big number of thermocouples are necessary, and their total thermal resistance must be equal to the thermal resistance of the heat sink. The reduced thermal resistance of the TEG decreases very much the temperature difference [62].

The thermal resistance of the heat sink is:

$$R_{\text{hs}} = \frac{T_{\text{heat sink}} - T_{\text{amb}}}{\dot{Q}_{\text{h}}} \tag{35}$$

where $\dot{Q}_{\text{h}}$ the heat flow is given by Eq. (18) through the TEG, $T_{heat\ sink}$ is the heat sink temperature, and $T_{\text{amb}}$ is the environmental temperature.

The thermal resistance and thermal conductance of TEG are linked with each other by an inverse ratio as:

$$R = \frac{1}{K} \tag{36}$$

The thermal energy through the TEG is written as:

$$\dot{Q}_h = \frac{T_{\text{heat source}} - T_{\text{amb}}}{R_{\text{tot}}} \tag{37}$$

considering that $R_{\text{hs}}$ and $R$ are connected in series with the total resistance $R_{\text{tot}} = R + R_{\text{hs}}$, and $T_{\text{heat source}}$ is the heat source temperature.

The following cases may be considered:

- If $\frac{R}{R_{\text{hs}}} < 1$, a big heat source-to-environment $\Delta T$ occurs across the heat sink and $\Delta T_{\text{TEG}} < \Delta T_{\text{heat sink}}$. In this case, a reduction of the thermoelectric conversion efficiency $\eta_{\text{TEG}}$ is observed, leading to a reduction of the TEG electric output power.

- If $\frac{R}{R_{\text{hs}}} > 1$, a big heat source-to-environment $\Delta T$ occurs across the TEG and $\Delta T_{\text{TEG}} > \Delta T_{\text{heat sink}}$. In this case, an increment of the $\eta_{\text{TEG}}$ is observed, leading to a limited electric output power.

- When $\frac{R}{R_{hs}} = 1$, the electric output power has a peak. In this case, the lengths and cross sections of thermoelectric legs are adapted, $R = R_{hs}$ and $\Delta T$ is equally divided between the heat sink and the TEG.

Also, when the heat sinks are attached to both sides of the TEG, the total thermal resistance (thermal interfaces resistances and thermal resistances of the heat sinks) is equal to the $R_{TEG}$ for maximum electric output power [62]. The contact resistance decreases the electric output power by decreasing $\Delta T$ across the TEG. Furthermore, the thermal contact resistance between the TEG and the heat sink or heat source is decreased to reduce the contact effect [63].

Astrain et al. [64] demonstrated the significance of decreasing the thermal resistance between the heat source and the hot side of the TEG, as well as the cold side of TEG and the environment. The numerical model assesses the TEG performance, taking into account the heat exchangers attached on both sides of the TEG, the heat source, as well as the heat sink. The results obtained show a good accuracy of the model. The results demonstrated that increasing by 10% the thermal resistances of both heat exchangers, the electric output power is improved by 8%. Martínez et al. [65] optimised the heat exchangers fixed on both sides of a TEG to maximise the electric output power. They have concluded that the thermal resistances of the heat exchangers are very important for TEG design. Zhou et al. [66] studied the heat transfer features of a TEG device. The heat transfer intensification on the cold side of the TEG leads to a significant reduction of the temperature and thermal resistance on this side, and implicitly a rise of the electric output power of the TEG device. Furthermore, Zhou et al. [66] highlighted that the refrigerant which flows by heat exchangers produce higher net powers than conventional heat sink with fins. An in-depth review of the heat sink for TEG and parameters affecting TEG performance is presented in [26].

The refrigeration system of the TEG has been assessed by Aranguren et al. [30]. This system consists of a multi-channel heat exchanger attached to the cold side of the TEG, another heat exchanger used to decrease the refrigerant temperature, the pump to circulate the refrigerant, and the connecting pipes. A numerical model has been implemented to compute the total thermal resistance and the power consumption in the system components. In this model, all system elements have been included to obtain an accurate analysis. The combination of computational and experimental results shows that the system configuration leading to the maximum net power is different with respect to the configuration resulting in the lowest total thermal resistance.

## 3. Applications using thermoelectrics in the power generation mode

The favourable characteristics of the thermoelectric devices promote the development of standalone TEGs for energy harvesting in a wide range of applications (**Figure 8**) as military, aerospace (e.g., powering spacecraft), biological systems (e.g., to power implanted pacemakers) and other applications (e.g., power for wristwatches or mobile communications) [67]. The key element to improve the energy conversion efficiency of TEG is the effect of waste heat recovery. Waste heat represents the heat produced by machines (e.g., exhaust pipes from automobiles), industrial processes (e.g., cooling towers, burnt solid waste and radioactive wastes), electrical equipment (e.g., kerosene lamps) and the human body. For various TEG applications (e.g., waste thermal power recovery using TEGs and powering of

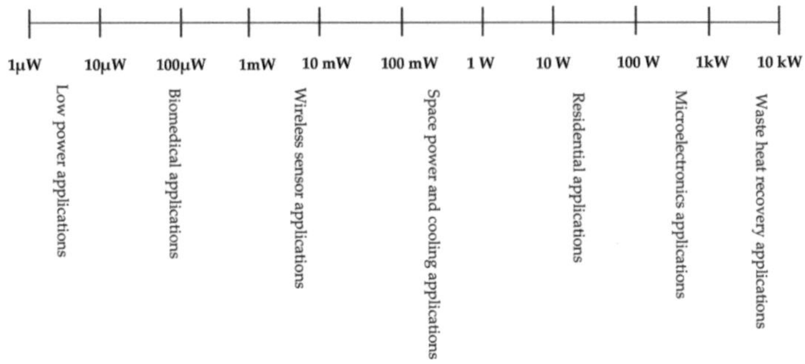


**Figure 8.**
*Energy conversion applications.*

wireless sensors by TEGs) even if $\Delta T$ is restricted, the available heat is higher than the capacity of the harvesters. In this case, the heat source delivers a constant heat flow rate at a constant $\Delta T$. Low $\eta_{\text{TEGmax}}$ in such applications does not mean low TEG performance [32].

## 3.1 Low-power generation for thermoelectric harvesting

### *3.1.1 Microelectronic applications*

The TEG devices are especially suitable for waste heat harvesting for low-power generation to supply electric energy for microelectronic applications. Wearable TEGs harvest heat generated by the body to generate electricity. For this reason, it is possible to use waste human body heat to power a TEG watch device. In this case, the wristwatch can capture the thermoelectric energy. Now, body-attached TEGs are commercially available products including watches operated by body temperature and thin film devices. Some manufacturers produce and commercialise wristwatches with an efficiency about 0.1% at 300 mV open circuit voltage from 1.5 K temperature drop and 22 μW of electric output power under of TEG normal operation. A thermo-clock wristwatch produces a voltage of 640 mV and gives a power of 13.8 μW for each °C of temperature difference. A wristwatch with 1040 thermoelements generates in the same conditions at about 200 mV [25]. The wearable TEG performance is affected by the utilisation of the free air convection cooling on the cold side of TEG, the low operating temperature difference between the body and environment, as well as the demand for systems that are thin and lightweight, being practical for long-term usage [68].

Furthermore, various microelectronic devices, like wireless sensor networks, mobile devices (e.g., mp3 player, smartphones and iPod), and biomedical devices are developed. The thermoelectric energy harvesters are microelectronic devices made of inorganic thermoelectric materials, at different dimensions, with a lifetime of about 5 years [69] and electric output powers are cardiac pacemakers ($P = 70 \div 100\ \mu W$) [70], pulse oximeter ($P = 100\ \mu W$) [71], wireless communication ($P \sim 3$ mW) [72], electrocardiography (ECG)/electroencephalography (EEG)/electromyography (EMG) with $P = 60 \div 200\ \mu W$ [73], EEG headband ($P = 2 \div 2.5$ mW) [74], ECG system ($P \sim 0.5\ \mu W$) [75], Hearing aid ($P \sim 1\ \mu W$) [76] and Wireless EEG [77]. Together with the progress of flexible thermoelectric materials (both organic and inorganic materials), flexible $\text{TEG}_\text{S}$ system benefits from special attention. The flexible thermoelectric materials and maximum electric output power of various TEG systems are reported in [44].

For these microelectronic devices, standard batteries are used. These batteries are made of various inorganic materials (like nickel, zinc, lithium, lead, mercury, sulphuric acid and cadmium) that are not friendly for the human body. In this case, the body-attached TEGs could be an alternative solution because the materials used are non-toxic [36].

A TEG to be applied in a network of body sensors has been presented in [78]. In this case, the device has been fixed in a body zone, where the maximum body heat has been obtained and also maximum energy. This equipment is capable of storing about 100 μW on the battery, leading to an output voltage of 2.4 V. Another TEG has been designed to be used on the wrist [79]. The output voltage of the device was 150 mV under normal conditions and an electric output power of 0.3 nW.

#### *3.1.2 TEG as a thermal energy sensor*

Thermal energy sensors (like heat-flux sensors, infrared sensors, power ultra-sound effect sensors, fluid-flow sensors and water condensation detectors) are used to convert heat flow rates into electrical signals by a TEG system [36].

The heat-flux sensors are used to evaluate the thermoelectric properties of micro-TEGs. In this case, the generated power and the thermoelectric conversion efficiency are measured with high accuracy [80]. The electrical signal generated by the heat-flux sensor is proportional to the heat flow rate applied to the sensor surface. The convective heat flow rate is measured from the temperature difference between two sides of a thermal resistive element plate placed across the flow of heat. The heat radiated from the mass is absorbed by the infrared sensor (IR) and the temperature increase leads to the generation of the Seebeck voltage. The thermoelectric IR sensor operates in a range from 7 to 14 μm [69].

### 3.2 High-power generation for thermoelectric harvesting

About 70% of energy in the world is wasted as heat and is released into the environment with a significant influence on global warming [81]. The waste heat energy released into the environment is one of the most significant sources of clean, fuel-free and cheap energy available. The unfavourable effects of global warming can be diminished using the TEG system by harvesting waste heat from residential, industrial and commercial fields [36].

TEG is substantially used to recover waste heat in different applications ranging from μW to MW. Different waste heat sources and temperature ranges for thermoelectric energy harvesting are shown in **Table 2** [69].

#### *3.2.1 Automotive applications*

The automotive industry is considered as the most attractive sector in which TEGs are used to recover the lost heat. Various leading automobile manufacturers develop TEGs ($P \sim 1$ kW) for waste recovery to reduce the costs of the fuel for their vehicles [82]. It has been demonstrated that vehicles (the gasoline vehicle and hybrid electric vehicles) have inefficient internal combustion engines. This can be observed in the Sankey diagram depicted in [83], which presents the energy flow direction of an internal combustion engine. The fuel combustion is used in a proportion of 25% for vehicle operation, 30% is lost into the coolant and 40% is lost as waste heat with exhaust gases. In this case, the TEG technology could be an option to recuperate the waste heat energy for gasoline vehicles and hybrid electric vehicles. A significant power conversion could be achieved by combining cooling system losses with the heat recovery from automobile exhausts. The use of TEG systems with an energy conversion of 5% would raise the electrical energy in a vehicle by 6% (5% from exhaust gases and 1% from the cooling system) [25].

| Temperature ranges, °C | Temperature, °C | Waste heat sources |
|---|---|---|
| High temperature (>650°C) | 650–760 | Aluminium refining furnaces |
| | 760–815 | Copper reverberatory furnace |
| | 760–110 | Copper refining furnace |
| | 620–730 | Cement kiln |
| | | Hydrogen plants |
| Medium temperature (230–650°C) | 315–600 | Reciprocating engine exhausts |
| | 425–650 | Catalytic crackers |
| | 425–650 | Annealing furnace cooling systems |
| Low temperature (230–650°C) | 32–55 | Cooling water |
| | 27–50 | Air compressors |
| | 27–88 | Forming Dies and pumps |

**Table 2.**
*Different waste heat sources and temperature ranges for thermoelectric harvesting technology.*

A TEG with $ZT = 1.25$ and efficiency of 10%, can recover about 35–40% of the power from the exhaust gas where the power generated can help to increase the efficiency to up 16% [84]. The components where TEGs could be attached in a vehicle are the exhaust system and the radiators. In this case, the amount of waste heat is decreased and exhaust temperatures are reduced. These aspects require more efficiency from the TEG device. Furthermore, the design of such power conversion system takes into account various heat exchangers mounted on the TEG device. These systems have a lifecycle from 10 to 30 years and the materials accumulated on their surfaces from the exhaust gas, air or coolant represent a major concern in order to not damage their proper operation [85]. Important testing is helpful to confirm the reliability of TEG systems in automotive applications. Furthermore, the design requires knowing the maximum electric output power and conversion efficiency from TEG systems [37].

The main components of the automotive TEG that considers waste heat like their energy source are one heat exchanger which takes heat from engine coolant and the exhaust gases and release it to the hot side of the TEG; the TEG system; one heat exchanger which takes the heat from the TEG and releases it to the coolant or to the air; the electrical power conditioning and the interface unit to supply the electric output power of the TEG system to the automobile electric system (**Figure 9**). Supplementary at these components, there are secondary components (e.g., the electronic unit, the electric pump, sensors system, valves, fans and so on) depending on the vehicle design and application type [85].

Thacher et al. [86] carried out the feasibility of the TEG system installed in the exhaust pipe in a light truck by connecting a series of 16 TEG modules. The

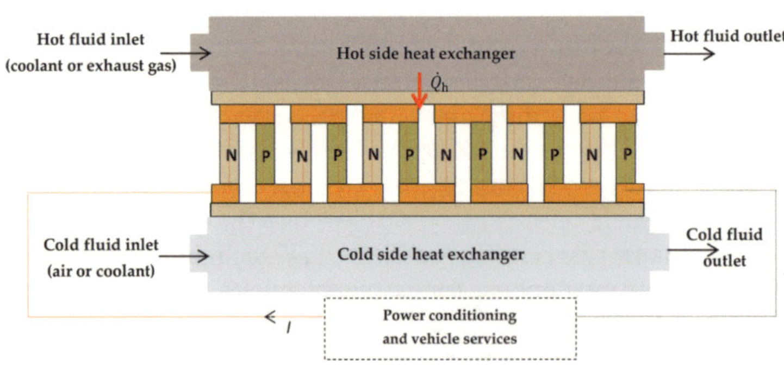


**Figure 9.**
*The main components of an automotive TEG system.*

experimental results showed good performance of the system at high speeds. Hsiao et al. [87] carried out an analytical and experimental assessment of the waste heat recovery system from an automobile engine. The results showed better performance by attaching TEGs to the exhaust pipe than to the radiators. Hsu et al. [88] introduced a heat exchanger with 8 TEGs and 8 air-cooled heat sink assemblies, obtaining a maximum power of 44 W. An application to recover waste heat has been developed by Hsu et al. [89], for a system consisting of 24 TEGs used to convert heat from the exhaust pipe of a vehicle to electrical energy. The results show a temperature increase at the hot side $T_h$ from 323 to 403 K and a load resistance of 23–30 Ω to harvest the waste heat for the system. Tian et al. [90] theoretically analysed the performance between a segmented TEG ($Bi_2Te_3$ used in low-temperature region and Skutterudite in high-temperature areas) used to recover exhaust waste heat from a diesel engine and traditional TEG. They found that a segmented TEG is suitable for large temperature difference and a high-temperature heat source, and has a higher potential for waste heat recovery compared to the traditional device. Meng et al. [91] addressed the automobile performance when applying TEG in exhaust waste heat recovery. The results showed that the effects of the different properties and the heat loss to the environmental gas on performance are considerable.

The conversion efficiency for the TEG system could be in the range of 5–10% [83]. The researchers' attention is focused on the development of new thermoelectric materials that offer improved energy conversion efficiency and a working temperature range more significant than for internal combustion engines. It is planned by 2020, about 90% of cars in the USA to have mounted TEGs for their cooling equipment, thus replacing the air conditioning systems. In this case, an amount of 5% of daily average gasoline consumption would be saved and a significant reduction of greenhouse gas emissions would be obtained [25].

To recover waste heat from the exhaust gas of engines, the research efforts of manufacturers focused on different solutions to compete in the production of ever-cleaner cars. Even if the cost of the bismuth telluride is relatively high, the technical feasibility of TEGs for the automobile industry is widely demonstrated, making it very attractive. The goal of the manufacturers is to develop TEG systems with automated production and low-cost thermoelectric materials [29].

### *3.2.2 Air applications*

#### *3.2.2.1 Space vehicle applications*

A considerable amount of heat is released into the atmosphere from space vehicles (turbine engines from helicopters and aircraft jet engines) [29]. To obtain a significant reduction of the gas pollutant into the environment, it is necessary a remarkable reduction of electricity consumption and utilisation of the available energy in these types of vehicles. Implicitly, their operating costs are reduced [25].

To power these space vehicles, TEG systems are used (e.g., on fixed-wing aircraft). The backup TEG is a type of static thermoelectric energy harvesting system with a significant temperature difference across the TEG around 100°C [92].

TEG for energy harvesting uses the available temperature gradient and collects sufficient energy to power up an energy wireless sensor node (WSN) to be autonomous. This WSN is used for health monitoring systems (HMS) in an aircraft structure. The main components of a WSN are the energy source and the wireless sensor unit. An in-depth review of WSN mechanisms and applications is presented in [10].

A TEG energy harvesting captures enough energy for a wireless sensor. One side of the TEG is fixed directly to the fuselage and the other side is attached to a phase-change material (PCM) heat storage unit to obtain a temperature difference during take-off and landing (**Figure 10**). PCM is considered an essential element for the heat storage unit because it can maximise the $\Delta T$ of the TEG system to solve the low TEG conversion efficiency [93]. In this case, the electrical energy is generated [94]. Water is an adequate PCM for heat storage. The temperature difference across the TEG is obtained from the slow changing temperature of the heat storage unit and the rapidly changing temperature of the aircraft fuselage. A lot of energy is produced during the PC, through latent heat [95, 96].

An application of $Bi_2Te_3$ modules on turbine nozzles has been addressed in [97]. Even though the electric power that can be harvested may be significant, the weight of the cold exchanger is still excessive for the specific application.

Future applications in aircraft may be envisioned in locations in which there are hot and cold heat flows, especially with the use of light thermoelectric materials. However, one of the main issues remains the weight of the heat exchangers [29].

#### *3.2.2.2 Spacecraft applications*

The radioisotope thermoelectric generators (RTGs) are a solid and highly reliable source of electrical energy to power space vehicles being capable of operating in vacuum and to resist at high vibrations [98, 99]. RTGs are used to power space vehicles for distant NASA space expeditions (e.g., several years or several decades) where sunlight is not enough to supply solar panels [29]. The natural radioactive decay of plutonium-238 releases huge amounts of heat, which is suitable for utilisation in RTGs to convert it into electricity. The thermoelectric materials used the thermocouples of the RTGs are adequate for high temperature considering that the heat source temperature is about 1000°C [100]. These semiconductor materials can be silicon germanium (Si Ge), lead tin telluride (PbSnTe), tellurides of antimony, germanium, and silver (TAGS) and lead telluride (PbTe).

### *3.2.3 Marine applications*

Up to now, just a few surveys have been performed in the marine industry due to the lack of clear and stringent international rules at the global level. The marine transport has a significant influence on climate change because is a large amount of the greenhouse gas emissions [29]. The naval transport generates a wide amount of waste heat, used to provide thermal energy onboard and seldom electrical energy.

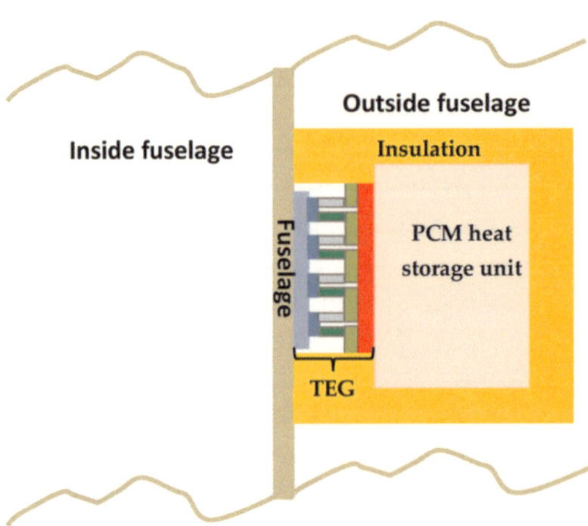


**Figure 10.**
*Schematic of the thermoelectric harvesting system in an aircraft.*

The heat sources on the marine vessels are the main engine, lubrication oil cooler, an electrical generating unit, generator and incinerators. The utilisation of waste heat onboard is for heating heavy fuel oil and accommodation places, and for freshwater production. The main engine represents the principal source of waste heat. Board incinerators are used for burning the onboard waste instead to be thrown overboard to pollute the sea water. The incinerators are the most favourable TEG systems due to the availability of their high-temperature differences [101, 102].

The specialists' attention is focused on the future design and optimisation of the high-power density TEGs for the marine environment, as well as on the development of hybrid thermoelectric ships considered as green platforms for assessing the efficiency of TEGs [103].

#### *3.2.4 Industrial applications*

The industry is the field where most amounts of heat are emitted and released into the atmosphere in the form of flue gases and radiant heat energy with a negative impact to the environmental pollution (emissions of $CO_2$). For this reason, thermoelectric harvesters are good candidates to recover waste heat from industries and convert it into useful power (e.g., to supply small sensing electronic device in a plant).

Utilisation of TEGs in the industrial field is beneficial from two points of view:

- in the industrial applications where recoverability of the waste heat by the conventional system (radiated heat energy) is very difficult to be done;
- in the industrial applications where the use of thermoelectric materials reduces the need for maintenance of the systems and the price of the electric power is low, even if the efficiency is low [29].

The results of a test carried out on a TEG system attached at a carburising furnace (made of 16 $Bi_2Te_3$ modules and a heat exchanger) are indicated in [104]. The system harvested about 20% of the heat ($P$ = 4 kW). The maximum electrical output power generated by TEG has been approximately 214 W, leading to thermoelectric conversion efficiency 5%. Aranguren et al. [105] built a TEG prototype. The TEG has been attached at the exhaust of a combustion chamber, with 48 modules connected in series and two different kinds of finned heat sinks, heat exchangers and heat pipes. This TEG was used to recover waste heat from the combustion chamber. In this case, the main objective has been to maximise the electric output power generated by the TEG.

For this reason, the dissipation systems have been used on both sides of TEG. This prototype has obtained a 21.56 W of net power using about 100 W/m$^2$ from the exhaust gases of the combustion chamber. To recover the radiant heat from melted metal from the steelmaking industry, the TEG systems are also considered good candidates [29].

Furthermore, TEGs are useful for recovery of waste heat from the cement rotary kiln to generate electricity, considering that the rotary kiln is the main equipment used for large-scale industrial cement production [106]. The performance of this hybrid $Bi_2Te_3$ and PbTe thermoelectric heat recovery system is obtained by developing a mathematical model. In this case, about 211 kW electrical output power and 3283 kW heat loss are saved by using a thermoelectric waste heat energy recovery system. The contribution of TEG is about 2%.

The electric output power evaluation of a TEG system attached to an industrial thermal oil heater is presented in Barma et al. [107]. The impact of different design and flow parameters has been assessed to maximise the electrical output power. The estimated annual electrical power generation from the proposed system was about 181,209 kWh. The thermal efficiency of the TEG based on recently developed thermoelectric materials (N-type hot forged $Bi_2Te_3$ and P-type $(Bi,Sb)_2Te_3$ used for the temperature range of 300–573 K) was enhanced up to 8.18%.

### 3.2.5 Residential applications

In residential applications, TEGs can be feasible where the heat is transferred from high temperatures to a reduced temperature heat source, and then is released into the environment. In addition, TEG can be also feasible when thermal energy is accessible in high amounts without additional costs. Types of residential applications where TEG systems could be mounted are with TEGs attached to domestic boilers, TEGs attached to stoves, as well as TEGs attached to solar systems [27].

#### 3.2.5.1 TEGs connected to the local heating boiler

The heating boilers for residential applications provide central heating and hot water. These heating boilers are highly used in the places where the winter season with temperatures under 0°C has a long duration and the heating is necessary for all this period. The fuels used by these heating boilers can be biomass (e.g., firewood, wood pellets, wood chips) or renewable resources that provide a lower carbon footprint compared to fossil fuels [108, 109]. In spite of higher pollution, some residential applications use fossil fuel-fired boilers (boilers supplied with natural gas) due to their low maintenance [27]. Furthermore, the fuel combustion is high, providing combustion temperatures bigger than 1000°C.

Instead, the heating boilers contain enhanced heat transfer surfaces due to the fins, and inside them combustion takes place at over 500°C. The heat obtained is used for water heating at temperatures fewer than 80°C. By considering that some thermoelectric materials are available for high temperatures, TEGs are very suitable for this type of equipment. In this case, high-temperature differences are obtained if the TEG cold side is mounted to the water heating side and the TEG hot side is mounted to the combustion chamber of the heat exchanger. The TEG system attached to the heating boiler must provide an electric output power $P = 30 \div 70$W, as this boiler has to generate the power necessary to supply the auxiliary devices and the pumps of the heating system. These boilers are widely reviewed in [27].

#### 3.2.5.2 TEGs attached to the stove

At the global level, over than 14% of the population is still living without electricity access according to the Energy Access Outlook 2017 [110]. To deliver a small amount of electricity by using power plants to this population could be very costly. Grid connection of villages in remote areas supposes to take into account the cost of the connection of the new power lines to the grid and the distribution cost on long distances [29]. Also, traditional biomass stoves ('threestone fires', 'built-in stoves' or 'mud-stoves') have a reduced thermal efficiency due to incomplete fuel burning, and a lot of the heat generated is wasted through the exhausts. Furthermore, the indoor air quality is very poor related to the utilisation of biomass fuels with a negative impact on the users' health (e.g., lung diseases, respiratory tract infections, cardiac problems, stroke, eye diseases, tuberculosis and cancer) [111].

Considering that most of the stoves are used in the rural areas in remote locations away from the grid, where the income of the population is very low, the solutions with such hybrid systems (TEGs attached to stoves) must be as inexpensive as possible. These hybrid systems must be very reliable and durable, taking into account this vital problem [111]. In this case, a TEG device attached to the stove equipment used for heating and food preparation could be an attractive option. The biomass stoves-powered TEG uses heat exchangers on both sides of TEG, as well as a power management system. The internal temperatures of a stove are bigger than 600°C, while most commercial TEGs can operate continuously at temperatures higher than 250°C, so that an appropriate hot side temperature could be obtained. To obtain a maximum electric output power, the cold side temperature must be very low. In this case, the cold sink dissipates a big amount of heat maintaining its low temperature [112].

According to the literature survey, various extensive reviews about stove-attached TEG systems have been presented [27, 29, 111–114]. A good option is the utilisation of the stoves using water as the cooling medium of the TEG to produce the maximum electric output power. The electrical output power decreases if the number of the TEGs is increased on the same heat sink. Such a hybrid system is suitable for complete combustion. The overall efficiency could be substantially improved [111].

#### *3.2.5.3 TEGs attached to the solar systems*

Solar TEG (STEG) systems, PV systems and concentrating solar power plants can generate electricity by using the solar heat. A STEG is composed of a TEG system sandwiched between a solar absorber and a heat sink as shown in **Figure 11**. The solar flux is absorbed by the solar absorber and concentrated into one point. Then, the heat is transferred through TEG by using a pipeline, and is partially

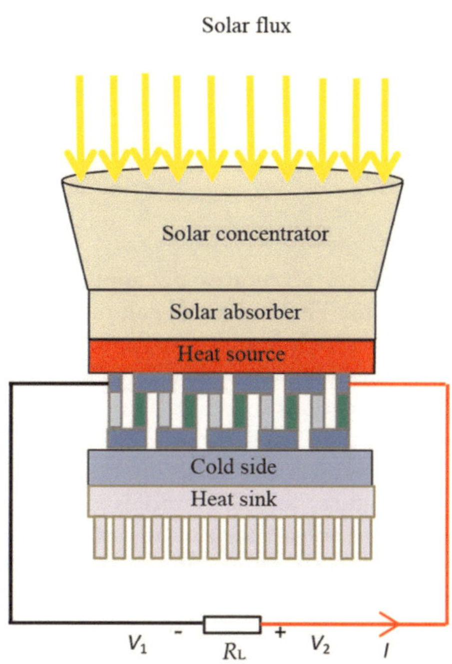


**Figure 11.**
*Components of STEG system.*

converted into electrical power by the TEG. A heat sink rejects the excess heat at the cold junction of the TEG to keep a proper $\Delta T$ across the TEG [115].

Due to the development of the thermoelectric materials, a solar TEG with an incident flux of 100 kW/m$^2$ and a hot side temperature of 1000°C could obtain 15.9% conversion efficiency. The solar TEG is very attractive for standalone power conversion. The efficiency of a solar TEG depends on both the efficiency with which sunlight is absorbed and converted into heat, and the TEG efficiency $\eta_{TEG}$. Furthermore, the total efficiency of a solar TEG is also influenced by the heat lost from the surface. The efficiency of solar TEG systems is relatively small due to the low Carnot efficiency provoked by the reduced temperature difference across the TEG and the reduced $ZT$ [116]. Its improvement needs to rise temperature differences and to develop new materials with high $ZT$ like nanostructured and complex bulk materials (e.g., a device with $ZT = 2$ and a temperature of 1500°C would lead to obtain a conversion efficiency about 30.6%) [117]. According to the literature survey, both residential and commercial applications gain much more interest in the regions of incident solar radiation of solar TEGs. This can be explained by the fact that most of the heat released at the cold side of the TEG can be used for domestic hot water and space heating [115].

### *3.2.6 Grid integration of TEG*

Most TEG applications have been designed for autonomous operation within a local system. Of course, the TEG output may be connected to different types of loads. In general, a TEG can be seen as a renewable energy power generation source that supplies an autonomous system or a grid-connected system. To be suitable for grid connection, the TEG needs an appropriate power conditioning system. This power conditioning system has to be a power electronic system, with specific regulation capabilities, different with respect to the ones used for solar photovoltaic and wind power systems [114], because the TEG operating conditions are different with respect to the other renewable energy sources. Molina et al. [118] proposed a control strategy to perform energy conversion from DC to AC output voltage, which maintains the operation of the thermoelectric device at the MPP. In the same proposal, active and reactive power controls are addressed by using a dedicated power conditioning system.

## 4. Conclusions

This chapter has addressed the structures and applications of TEGs in various contexts. It has emerged that the TEG is a viable solution for energy harvesting, able to supply electrical loads in relatively low-power applications. The TEG efficiency is also typically low. Thereby, the advantages of using TEG have to be found in the characteristics of specific applications in which there is a significantly high-temperature difference across the TEG system, and other solutions with higher efficiency cannot be applied because of various limitations. These limitations may be the relatively high temperatures for the materials adopted, the strict requirements on the system to be used (regarding the type of operation, emissions of pollutants, the position of the device during operation or noise). In these cases, TEGs may be fully competitive with the other solutions.

In particular, the use of TEGs is entirely consistent with the provision of green energy through energy harvesting from even small temperature differences. Some low-power applications have been identified on electronic circuits, sensors, waste heat recovery, residential energy harvesting and automotive systems. In other applications to enhance the efficiency of the systems for energy production with a higher power, the efficiency increase is still somewhat limited to consider that an investment in TEG integration may be profitable. Nevertheless, there is a growing interest in the potential of thermoelectric applications. For the future, faster development of TEG solutions can be expected in a broader range of green energy applications. This development depends on improvements in the TEG technology, better information on the TEG characteristics, and the testing of new solutions aimed at promoting better integration in the energy production systems.

## Nomenclature

### Acronyms

| | |
|---|---|
| *COP* | coefficient of performance |
| DC | direct current |
| ECG | electrocardiography |
| EEG | electroencephalography |
| EMG | electromyography |
| HMS | health monitoring systems |
| IR | infrared sensor |
| MPP | maximum power point |
| MPPT | maximum power point tracking |
| PANI | polyaniline |
| PA | polyamide |
| PCM | phase-change material |
| PC | phase-change |
| PEDOT | poly |
| RF | radio-frequency |
| RTG | radioisotope thermoelectric generator |
| STEG | solar thermoelectric generator |
| TEG | thermoelectric generator |
| WSN | wireless sensor node |

### Symbols

| | |
|---|---|
| $E$ | electric field ($V \cdot m^{-1}$) |
| $I$ | electrical current (A) |
| $J$ | current density ($A \cdot m^{-2}$) |
| $\bar{J}$ | relative current density ($A \cdot m^{-2}$) |
| $k$ | thermal conductivity ($W \cdot m^{-1} \cdot K^{-1}$) |
| $K$ | thermal conductance ($W \cdot K^{-1}$) |
| $L$ | length (m) |
| $m$ | resistance ratio |
| $n$ | number of thermoelectric couples |
| $P$ | electric power (W) |
| $\dot{Q}$ | heat flow rate (W) |
| $R$ | resistance (Ω) |
| $S$ | cross-section ($m^2$) |
| $t$ | temperature (°C) |
| $T$ | absolute temperature (K) |
| $\Delta T$ | temperature difference (K) |

| | |
|---|---|
| $\nabla T$ | temperature gradient (K) |
| $u$ | compatibility factor |
| $V$ | voltage (V) |
| $x$ | slenderness ratio |
| $y$ | external load parameter |
| $Z$ | thermoelectric figure-of-merit ($W \cdot K^{-1}$) |
| $ZT$ | dimensionless figure-of-merit |

### Greek symbols

| | |
|---|---|
| $\alpha$ | Seebeck coefficient ($V \cdot K^{-1}$) |
| $\varepsilon$ | fraction of waste heat passed through the thermoelectric couples |
| $\eta$ | (efficiency) |
| $\mu$ | Thomson coefficient ($V \cdot K^{-1}$) |
| $\pi$ | Peltier coefficient (V) |
| $\rho$ | electrical resistivity ($\Omega \cdot m$) |
| $\sigma$ | electrical conductivity ($S \cdot m^{-1}$) |

### Subscripts

| | |
|---|---|
| a | contact |
| amb | (ambient) |
| c | cold side |
| C | Carnot |
| h | hot side |
| hs | heat sink |
| L | load |
| N | N-type semiconductor |
| P | P-type semiconductor |
| max | maximum |
| r | reduced |
| t | thermal |
| tot | total |

## Author details

Diana Enescu
Department of Electronics, Telecommunications and Energy, Valahia University of Targoviste, Targoviste, Romania

*Address all correspondence to: diana.enescu@valahia.ro

## References

[1] Roundy S, Steingart D, Frechette L, Wright P, Rabaey J. Power sources for wireless sensor networks. In: 1st European Workshop on Wireless Sensor Networks Berlin; 2004

[2] Vullers RJM, van Schaijk R, Doms I, Van Hoof C, Mertens R. Micropower energy harvesting. Solid-State Electronics. 2009;**53**(7):684-693. DOI: 10.1016/j.sse.2008.12.011

[3] Steingart D, Roundy S, Wright PK, Evans JW. Micropower materials development for wireless sensor networks. MRS Bulletin. 2008;**33**(4): 408-409. DOI: 10.1557/mrs2008.81

[4] Sim ZW. Radio frequency energy harvesting for embedded sensor networks in the natural environment [MS Degree thesis]. Manchester, England; 2011

[5] Snyder GJ. Thermoelectric energy harvesting. In: Priya S, Inman DJ, editors. Energy Harvesting Technologies. Boston, MA, USA: Springer; 2009. pp. 325-336. DOI: 10.1007/978-0-387-76464-1_11

[6] Cheng TC, Cheng CH, Huang ZZ, Liao GC. Development of an energy-saving module via combination of solar cells and thermoelectric coolers for green building applications. Energy. 2011;**36**(1):133-140. DOI: 10.1016/j.energy.2010.10.061

[7] Wang CC, Hung CI, Chen WH. Design of heat sink for improving the performance of thermoelectric generator using two-stage optimization. Energy. 2012;**39**:236-245. DOI: 10.1016/j.energy.2012.01.025

[8] Camacho-Medina P, Olivares-Robles PA, Vargas-Almeida A, Solorio-Ordaz A. Maximum power of thermally and electrically coupled thermoelectric generators. Entropy. 2014;**16**:2890-2903. DOI: 10.3390/e16052890

[9] Brownell E, Hodes M. Optimal design of thermoelectric generators embedded in a thermal resistance network. IEEE Transactions on Components, Packaging and Manufacturing Technology. 2014;**4**(4): 612-621. DOI: 10.1109/TCPMT.2013.2295169

[10] Shaikh FK, Zeadally S. Energy harvesting in wireless sensor networks: A comprehensive review. Renewable and Sustainable Energy Reviews. 2017; **55**:1041-1054. DOI: 10.1016/j.rser.2015.11.010

[11] Simons RE, Ellsworth MJ, Chu RC. An assessment of module cooling enhancement with thermoelectric coolers. Journal of Heat Transfer-Transaction ASME. 2005;**127**:76-84. DOI: 10.1115/1.1852496

[12] Cheng TC, Cheng CH, Huang ZZ, Liao GC. Development of an energy-saving module via combination of solar cells and thermoelectric coolers for green building applications. Energy. 2011;**36**:133-140. DOI: 10.1016/j.energy.2010.10.061

[13] Enescu D. Thermoelctric refrigeration principle. In: Patricia Aranguren P, editor. Bringing Thermoelectricity into Reality, INTECH Publishing; 2018. pp. 221-246. DOI: 10.5772/intechopen.75439

[14] Kong LB, Li T, Hng HH, Boey F, Zhang T, Li S. Waste Energy Harvesting. Mechanical and Thermal Energies. Verlag Berlin Heidelberg, Germany: Springer; 2014. 592 p. DOI: 10.1007/978-3-642-54634-1

[15] Champier D, Bedecarrats JP, Kousksou T, Rivaletto M, Strub F, Pignolet P. Study of a TE (thermoelectric) generator incorporated in a multifunction wood stove. Energy.

2011;**36**:1518-1526. DOI: 10.1016/j.energy.2011.01.01

[16] Champier D, Bedecarrats JP, Rivaletto M, Strub F. Thermoelectric power generation from biomass cook stoves. Energy. 2010;**35**:935-942. DOI: 10.1016/j.energy.2009.07.015

[17] Marciá-Barber E. Thermoelectric Materials. Advances and Applications. NY, USA: Taylor & Francis Group, Pan Stanford. 2015. 350 p. ISBN 978-981-4463-53-9

[18] Neeli G, Behara DK, Kumar MK. State of the art review on thermoelectric materials. International Journal of Science and Research. 2016;**5**:1833-1844. ISSN:2319-7064

[19] Goupil C, Seifert W, Zabrocki K, Müller E, Snyder GF. Thermodynamics of thermoelectric phenomena and applications. Entropy. 2011;**13**: 1481-1517. DOI: 10.3390/e13081481

[20] Agrawal M. Basics of Irreversible Thermodynamics. Stanford, CA, USA: Stanford University; 2005

[21] DiSalvo FJ. Thermoelectric cooling and power generation. Science. 1999; **285**(5248):703-706. DOI: 10.1126/science.285.5428.703

[22] von Lukowicz M, Abbe E, Schmiel T, Tajmar M. Thermoelectric generators on satellites—An approach for waste heat recovery in space. Energies. 2016; **9**(7):1-14. DOI: 10.3390/en9070541

[23] Hodes M. Optimal pellet geometries for thermoelectric power generation. IEEE Transactions on Components and Packaging. 2010;**33**(2):307-318. DOI: 10.1109/TCAPT.2009.2039934

[24] Orr B, Akbarzadeh A, Mochizuki M, Singh R. A review of car waste heat recovery systems utilising thermoelectric generators and heat pipes. Applied Thermal Engineering. 2016;**101**:490-495. DOI: 10.1016/j.applthermaleng.2015.10.081

[25] Ando Junior OH, Maran ALO, Henao NC. A review of the development and applications of thermoelectric microgenerators for energy harvesting. Renewable and Sustainable Energy Reviews. 2018;**91**:376-393. DOI: 10.1016/j.rser.2018.03.052

[26] Elghool A, Basrawi F, Ibrahim TK, Habib K, Ibrahim H, Idris DMND. A review on heat sink for thermo-electric power generation: Classifications and parameters affecting performance. Energy Conversion and Management. 2017;**134**:260-277. DOI: 10.1016/j.enconman.2016.12.046

[27] Kütt L, Millar J, Karttunen A, Lehtonen M, Karppinen M. Thermoelectric applications for energy harvesting in domestic applications and micro-production units. Part I: Thermoelectric concepts, domestic boilers and biomass stoves. Renewable and Sustainable Energy Reviews. 2018;**98**: 519-544. DOI: 10.1016/j.rser.2017.03.051

[28] Thermoelectric power generation: Properties, application and novel TCAD simulation. In: Proceedings of the IEEE International Conference on Power Electronics and Applications; 30 August–1 September 2011. Birmingham, UK: IEEE; 2011. pp. 1-10

[29] Champier D. Thermoelectric generators: A review of applications. Energy Conversion and Management. 2017;**140**:167-181. DOI: 10.1016/j.enconman.2017.02.070

[30] Aranguren P, Astrain D, Pérez MG. Computational and experimental study of a complete heat dissipation system using water as heat carrier placed on a thermoelectric generator. Energy. 2014; **74**:346-358. DOI: 10.1016/j.energy.2014.06.094

[31] Rowe DM. Handbook of Thermoelectrics. Introduction.

Boca Raton, Fl, USA: CRC Press, Taylor & Francis Group; 1995. 720 p. ISBN: 9780849301469

[32] Kiziroglou ME, Yeatman EM. Materials and techniques for energy harvesting. In: Kilner JA, Skinner SJ, Irvine SJC, Edwards PP, editors. Woodhead Publishing Series in Energy, Functional Materials for Sustainable Energy Applications. Cambridge, UK: Woodhead Publishing; 2012. pp. 541-572. DOI: 10.1533/9780857096371.4.539

[33] Ali H, Sahin AZ, Yilbas BS. Thermodynamic analysis of a thermoelectric power generator in relation to geometric configuration device pins. Energy Conversion and Management. 2014;**78**:634-640. DOI: 10.1016/j.enconman.2013.11.029

[34] Crane DT, Jackson GS. Optimization of cross flow heat exchangers for thermoelectric waste heat recovery. Energy Conversion and Management. 2004;**45**(9–10):1565-1582. DOI: 10.1016/j.enconman.2003.09.003

[35] Yilbas BS, Sahin AZ. Thermoelectric device and optimum external load parameter and slenderness ratio. Energy. 2010;**35**(12):5380-5384. DOI: 10.1016/j.energy.2010.07.019

[36] Aswal DK, Basu R, Singh A. Key issues in development of thermoelectric power generators: High figure-of-merit materials and their highly conducting interfaces with metallic interconnects. Energy Conversion and Management. 2016;**114**:50-67. DOI: 10.1016/j.enconman.2016.01.065

[37] Niu X, Yu J, Wang S. Experimental study on low-temperature waste heat thermoelectric generator. Journal of Power Sources. 2009;**188**:621-626. DOI: 10.1016/j.jpowsour.2008.12.067

[38] Sahin AZ, Yilbas BS. The thermoelement as thermoelectric power generator: Effect of leg geometry on the efficiency and power generation. Energy Conversion and Management. 2013;**65**:26-32. DOI: 10.1016/j.enconman.2012.07.020

[39] Stobart R, Milner D. The potential for thermo-electric regeneration of energy in vehicles. SAE Technical Papers. 2009;**1**:1-14. DOI: 10.4271/2009-01-1333

[40] Rosenberg Y, Gelbstein Y, Dariel MP. Phase separation and thermoelectric properties of the Pb0.25Sn0.25Ge0.5Te compound. Journal of Alloys and Compounds. 2012;**526**:31-38. DOI: 10.1016/j.jallcom.2012.02.099

[41] Gelbstein Y, Davidow J. Highly efficient functional GexPb1-xTe based thermoelectric alloys. Physical Chemistry Chemical Physics. 2014;**16**:20120-20126. DOI: 10.1039/C4CP02399D

[42] Kirievsky K, Shlimovich M, Fuks D, Gelbstein Y. An ab-initio study of the thermoelectric enhancement potential in nano-grained TiNiSn. Physical Chemistry Chemical Physics. 2014;**16**:20023-20029. DOI: 10.1039/c4cp02868f

[43] Gelbstein Y, Tunbridge J, Dixon R, Reece MJ, Ning HP, Gilchrist R, et al. Physical, mechanical and structural properties of highly efficient nanostructured n- and p-silicides for practical thermoelectric applications. Journal of Electronic Materials. 2014;**43**:1703-1711. DOI: 10.1007/s11664-013-2848-9

[44] Zhou M, Al-Furjan MSH, Zou J, Liu W. A review on heat and mechanical energy harvesting from human—Principles, prototypes and perspectives. Renewable and Sustainable Energy Reviews. 2018;**82**:3582-3609. DOI: 10.1016/j.rser.2017.10.102

[45] Caillat T, Fleurial JP, Snyder GJ, Borshchevsky A. Development of high

efficiency segmented thermoelectric unicouples. In: Proceedings of the 20th International Conference on Thermoelectrics (ICT2001); 8-11 June 2001; Beijing, China: IEEE; 2001. pp. 282-285. DOI: 10.1109/ICT.2001.979888

[46] El-Genk MS, Saber HH. High efficiency segmented thermoelectric unicouple for operation between 973 and 300 K. Energy Conversion and Management. 2003;**44**(7):1069-1088. DOI: 10.1016/S0196-8904(02)00109-7

[47] Snyder GJ. Application of the compatibility factor to the design of segmented and cascaded thermoelectric generators. Applied Physics Letters. 2004;**84**(13):2436-2438

[48] Snyder GJ, Ursell T. Thermoelectric efficiency and compatibility. Physical Review Letters. 2003;**91**(14): 148301. DOI: 10.1103/PhysRevLett.91.148301

[49] Ngan PH, Christensen DV, Snyder GJ, et al. Towards high efficiency segmented thermoelectric unicouples. Physica Status Solidi. 2014;**211**(1):9-17. DOI: 10.1002/pssa.201330155

[50] Hung LT, Van Nong N, Han L, Bjørk R, Ngan PH, Holgate TC, et al. Segmented thermoelectric oxide-based module for high-temperature waste heat harvesting. Energy Technology. 2015;**3**:1143-1151. DOI: 10.1002/ente.201500176

[51] Ali H, Yilbas BS, Al-Sulaiman FA. Segmented thermoelectric generator: Influence of pin shape configuration on the device performance. Energy. 2016; **111**:439-452. DOI: 10.1016/j.energy.2016.06.003

[52] Ali H, Yilbas BS, Al-Sharafi A. Segmented thermoelectric generator: exponential area variation in leg. International Journal of Energy Research. 2018;**42**:477-489. DOI: 10.1002/er.3825

[53] Vikhor LN, Anatychuk LI. Generator modules of segmented thermoelements. Energy Conversion and Management. 2009;**50**:2366-2372. DOI: 10.1016/j.enconman.2009.05.020

[54] Zhang Q, Liao J, Tang Y, Gu M, Ming C, Qiu P, et al. Realizing a thermoelectric conversion efficiency of 12% in bismuth telluride/skutterudite segmented modules through full-parameter optimization and energy-loss minimized integration. Energy & Environmental Science. 2017;**10**: 956-963. DOI: 10.1039/C7EE00447H

[55] Zhang G, Fan L, Niu Z, Jiao K, Diao H, Du Q, et al. A comprehensive design method for segmented thermoelectric generator. Energy Conversion and Management. 2015;**106**: 510-519. DOI: 10.1016/j.enconman.2015.09.068

[56] Min G, Rowe DM. Optimization of thermoelectric module geometry for 'waste heat' electric power generation. Journal of Power Sources. 1992;**38**(3): 253-259. DOI: 10.1016/0378-7753(92)80114-Q

[57] Lavric ED. Sensitivity analysis of thermoelectric module performance with respect to geometry. Energy. 2010; **21**:133-138. DOI: 10.3303/CET1021023

[58] Meng JH, Zhang XX, Wang XD. Multi-objective and multi-parameter optimization of a thermoelectric generator module. Energy. 2014;**71**: 367-376. DOI: 10.1016/j.energy.2014.04.082

[59] Zhang G, Jiao K, Niu Z, Diao H, Du Q, Tian H, et al. Power and efficiency factors for comprehensive evaluation of thermoelectric generator materials. International Journal of Heat and Mass Transfer. 2016;**93**:1034-1037. DOI:

10.1016/j.ijheatmasstransfer.2015.10.051

[60] Jang JY, Tsai YC, Wu CW. A study of 3-D numerical simulation and comparison with experimental results on turbulent flow of venting flue gas using thermoelectric generator modules and plate fin heat sink. Energy. 2013;**53**: 270-281

[61] Borcuch M, Musiał M, Gumuła S, Sztekler K, Wojciechowski K. Analysis of the fins geometry of a hot-side heat exchanger on the performance parameters of a thermoelectric generation system. Applied Thermal Engineering. 2017;**127**:1355-1363. DOI: 10.1016/j.applthermaleng.2017.08.147

[62] Bierschenk JL. Optimized thermoelectrics for energy harvesting applications. In: Priya S, Inman DJ, editors. Energy Harvesting Technologies. USA: Springer; 2009. pp. 337-351. DOI: 10.1007/978-0-387-76464-1_12

[63] Liang G, Zhou J, Huang X. Analytical model of parallel thermoelectric generator. Applied Energy. 2011;**88**(12):5193-5199. DOI: 10.1016/j.apenergy.2011.07.041

[64] Astrain D, Vián JG, Martínez A, Rodríguez A. Study of the influence of heat exchangers' thermal resistances on a thermoelectric generation system. Energy. 2010;**35**(2):602-612. DOI: 10.1016/j.energy.2009.10.031

[65] Martínez A, Vián JG, Astrain D, Rodríguez A, Berrio I. Optimization of the heat exchangers of a thermoelectric generation system. Journal of Electronic Materials. 2010;**39**(9):1463-1468. DOI: 10.1007/s11664-010-1291-4

[66] Zhou ZG, Zhu DS, Wu HX, Zhang HS. Modeling, experimental study on the heat transfer characteristics of thermoelectric generator. Journal of Thermal Science. 2013;**22**(1):48-54. DOI: 10.1007/s11630-013-0591-4

[67] Riffat SB, Ma X. Thermoelectrics: A review of present and potential applications. Applied Thermal Engineering. 2003;**23**(8):913-935. DOI: 10.1016/S1359-4311(03)00012-7

[68] Settaluri KT, Lo H, Ram RJ. Thin thermoelectric generator system for body energy harvesting. Journal of Electronic Materials. 2012;**41**(6): 984-988. DOI: 10.1007/s11664-011-1834-3

[69] Zeb K, Ali SM, Khan B, Mehmood CA, Tareen N, Din W, et al. A survey on waste heat recovery: Electric power generation and potential prospects within Pakistan. Renewable and Sustainable Energy Reviews. 2017;**75**: 1142-1155. DOI: 10.1016/j.rser.2016.11.096

[70] Watkins C, Shen B, Venkatasubramanian R. Low-grade-heat energy harvesting using superlattice thermoelectrics for applications in implantable medical devices and sensors. In: Proceedings of the 24th International Conference on Thermoelectrics (ICT 2005); 19-23 June; Clemson, SC, USA: IEEE; 2005. pp. 265-267. DOI: 10.1109/ICT.2005.1519934

[71] Torfs T, Leonov V, Van Hoof C, Gyselinckx B. Body-heat powered autonomous pulse oximeter. In: Proceedings of the 5th IEEE Conference on Sensors; 22-25 October 2006; Daegu, South Korea: IEEE; 2007. pp. 427-430. DOI: 10.1109/ICSENS.2007.355497

[72] Mateu L, Codrea C, Lucas N, Pollak M, Spies P. Human body energy harvesting thermogenerator for sensing applications. In: Proceedings of the IEEE International Conference on Sensor Technologies and Applications (SENSORCOMM 2007); 14–20 October 2007; Valencia, Spain: IEEE; 2007.

pp. 366-372. DOI: 10.1109/SENSORCOMM.2007.4394949

[73] Zhang Y, Zhang F, Shakhsheer Y, Silver JD, Klinefelter A, Nagaraju M, et al. A batteryless 19 μW MICS/ISM-band energy harvesting body sensor node SoC for ExG applications. IEEE Journal of Solid-State Circuits. 2013; **48**(1):199-213. DOI: 10.1109/JSSC.2012.2221217

[74] Torfs T, Leonov V, Yazicioglu RF, Merken P, Van Hoof C, Vullers RJ, et al. Wearable autonomous wireless electro-encephalography system fully powered by human body heat. In: Proceedings of the IEEE International Conference on Sensors; 26-29 October 2008; Lecce, Italy: IEEE; 2008. pp. 1269-1272. DOI: 10.1109/ICSENS.2008.4716675

[75] Leonov V, Torfs T, Van Hoof C, Vullers RJ. Smart wireless sensors integrated in clothing: An electrocardiography system in a shirt powered using human body heat. Sensors & Transducers Journal. 2009; **107**(8):165. ISSN: 1726-5479

[76] Lay-Ekuakille A, Vendramin G, Trotta A, Mazzotta G. Thermoelectric generator design based on power from body heat for biomedical autonomous devices. In: Proceedings of IEEE international workshop on medical measurements and applications; 29-30 May 2009; Cetraro, Italy: IEEE; 2009. pp. 1-4. DOI: 10.1109/MEMEA.2009.5167942

[77] Carmo JP, Gonçalves LM, Correia JH. Thermoelectric microconverter for energy harvesting systems. IEEE Transactions on Industrial Electronics. 2010;**57**(3):861-867. DOI: 10.1109/TIE.2009.2034686

[78] Gyselinckx B, Van Hoof C, Ryckaert J, Yazicioglu RF, Fiorini P, Leonov V. Human++: Autonomous wireless sensors for body area networks. In: Custom Integrated Circuits Conference, San José, CA, 18–21 September 2005. IEEE; 2006. pp. 13-19

[79] Wang Z, Leonov V, Fiorini P, Van Hoof C. Realization of a wearable miniaturized thermoelectric generator for human body applications. Sensors and Actuators, A: Physical. 2009;**156**(1):95-102. DOI: 10.1016/j.sna.2009.02.028

[80] Beretta D, Massetti M, Lanzani G, Caironi M. Thermoelectric characterization of flexible micro-thermoelectric generators. The Review of Scientific Instruments. 2017;**88**(1):015103. DOI: 10.1063/1.4973417

[81] Zevenhovena R, Beyeneb A. The relative contribution of waste heat from power plants to global warming. Energy. 2011;**36**:3754-3762. DOI: 10.1016/j.energy.2010.10.010

[82] Elsheikh MM, Shnawah DA, Sabri MFM, Said SBM, Hassan MH, Bashir MBA, et al. A review on thermoelectric renewable energy: Principle parameters that affect their performance. Renewable and Sustainable Energy Reviews. 2014;**30**:337-355. DOI: 10.1016/j.rser.2013.10.027

[83] Yu C, Chau KT. Thermoelectric automotive waste heat energy recovery using maximum power point tracking. Energy Conversion and Management. 2009;**50**(6):1506-1512. DOI: 10.1016/j.enconman.2009.02.015

[84] Yu S, Du Q, Diao H, Shu G, Jiao K. Start-up modes of thermoelectric generator based on vehicle exhaust waste heat recovery. Applied Energy. 2015;**138**:276-290

[85] Yang J, Stabler FR. Automotive applications of thermoelectric materials. Journal of Electronic Materials. 2009; **38**(7):1245-1251. DOI: 10.1007/s11664-009-0680-z

[86] Thacher EF, Helenbrook BT, Karri KA, Richter CJ. Testing of an

automobile exhaust thermoelectric generator in a light truck. Proceedings of the Institution of Mechanical Engineers, Part D: Journal of Automobile Engineering. 2007;**221**(1): 95-107. DOI: 10.1243/ 09544070JAUTO51

[87] Hsiao YY, Chang WC, Chen SL. A mathematic model of thermoelectric module with applications on waste heat recovery from automobile engine. Energy. 2010;**35**(3):1447-1454. DOI: 10.1016/j.energy.2009.11.030

[88] Hsu CT, Yao DJ, Ye KJ, Yu B. Renewable energy of waste heat recovery system for automobiles. Journal of Renewable and Sustainable Energy. 2010;**2**:013105. DOI: 10.1063/ 1.3289832

[89] Hsu CT, Huang GH, Chu HS, Yu B, Yao DJ. Experiments and simulations on low-temperature waste heat harvesting system by thermoelectric power generators. Applied Energy. 2011;**88**(4): 1291-1297. DOI: 10.1016/j.apenergy. 2010.10.005

[90] Tian H, Sun X, Jia Q, Liang X, Shu G, Wang X. Comparison and parameter optimization of a segmented thermoelectric generator by using the high temperature exhaust of a diesel engine. Energy. 2015;**84**:121-130. DOI: 10.1016/j.energy.2015.02.063

[91] Meng JH, Wang XD, Chen WH. Performance investigation and design optimization of a thermoelectric generator applied in automobile exhaust waste heat recovery. Energy Conversion and Management. 2016;**120**:71-80. DOI: 10.1016/j.enconman.2016.04.080

[92] Janak L, Ancik Z, Vetiska J, Hadas Z. Thermoelectric generator based on MEMS module as an electric power backup in aerospace applications. Materials Today: Proceedings. 2015;**2**: 865-870. DOI: 10.1016/j.matpr.2015. 05.112

[93] Elefsiniotis A, Samson D, Becker T, Schmid U. Investigation of the performance of thermoelectric energy harvesters under real flight conditions. Journal of Electronic Materials. 2013; **42**(7):2301-2305. DOI: 10.1007/ s11664-012-2411-0

[94] Samson D, Otterpohl T, Kluge M, Schmid U, Becker T. Aircraft-specific thermoelectric generator module. Journal of Electronic Materials. 2010; **39**(9):2092-2095. DOI: 10.1007/ s11664-009-0997-7

[95] Samson D, Kluge M, Becker T, Schmid U. Energy harvesting for autonomous wireless sensor nodes in aircraft. Procedia Engineering. 2010;**5**: 1160-1163. DOI: 10.1016/j.proeng.2010. 09.317

[96] Elefsiniotis A, Kiziroglou ME, Wright SW, Toh TT, Mitcheson PD, Becker T, et al. Performance evaluation of a thermoelectric energy harvesting device using various phase change materials. Journal of Physics: Conference Series;**476**:1-5. DOI: 10.1088/1742-6596/476/1/012020

[97] Kousksou T, Bedecarrats J-P, Champier D, Pignolet P, Brillet C. Numerical study of thermoelectric power generation for an helicopter conical nozzle. Journal of Power Sources. 2011;**196**:4026-4032. DOI: 10.1016/j.jpowsour.2010.12.015

[98] Abelson RD. Space missions and applications. In: Rowe DM, editor. Thermoelectrics Handbook Macro to Nano. Boca Raton, Fl, USA: CRC Press, Taylor & Francis Group; 2006. pp. 56-1- 56-26. ISBN 0-8493-2264-2

[99] Fleurial JP, Caillat T, Nesmith BJ, Ewell RC, Woerner DF, Carr GC, et al. Thermoelectrics: From space power systems to terrestrial waste heat recovery applications. In: 2nd Thermoelectrics Applications Workshop. 2011

[100] LeBlanc S. Thermoelectric generators: Linking material properties and systems engineering for waste heat recovery applications. Sustainable Materials and Technologies. 2014;**1-2**: 26-35. DOI: 10.1016/j.susmat.2014.11.002

[101] Kristiansen NR, Nielsen HK. Potential for usage of thermoelectric generators on ships. Journal of Electronic Materials. 2010;**39**(9): 1746-1749. DOI: 10.1007/s11664-010-1189-1

[102] Kristiansen N, Snyder G, Nielsen H, Rosendahl L. Waste heat recovery from a marine waste incinerator using a thermoelectric generator. Journal of Electronic Materials. 2012;**41**(6): 1024-1029. DOI: 10.1007/s11664-012-2009-6

[103] Wallace TT. Development of marine thermoelectric heat recovery systems. In: The 2nd Thermoelectrics Applications Workshop 5 January 2011

[104] Kaibe H, Makino K, Kajihara T, Fujimoto S, Hachiuma H. Thermoelectric generating system attached to a carburizing furnace at Komatsu Ltd., Awazu Plant. AIP Conference Proceedings. 2012;**1449**: 524-527. DOI: 10.1063/1.4731609

[105] Aranguren P, Astrain D, Rodriguez A, Martinez A. Experimental investigation of the applicability of a thermoelectric generator to recover waste heat from a combustion chamber. Applied Energy. 2015;**152**:121-130. DOI: 10.1016/j.apenergy.2015.04.077

[106] Luo Q, Li P, Cai L, Zhou P, Tang D, Zhai P, et al. A thermoelectric waste-heat recovery system for Portland cement rotary kilns. Journal of Electronic Materials. 2015;**44**(6): 1750-1762. DOI: 10.1007/s11664-014-3543-1

[107] Barma MC, Riaz M, Saidur R, Long BD. Estimation of thermoelectric power generation by recovering waste heat from Biomass fired thermal oil heater. Energy Conversion and Management. 2015;**98**:303-313. DOI: 10.1016/j.enconman.2015.03.103

[108] Moser W, Friedl G, Haslinger W, Hofbauer H. Small-scale pellet boiler with thermoelectric generator. In: Proceedings of the 25th International Conference on Thermoelectrics. 6-10 August 2006; Vienna Austria: IEEE; 2007. pp. 349-353. DOI: 10.1109/ICT.2006.331221

[109] Thomson A, Liddell C. The suitability of wood pellet heating for domestic households: A review of literature. Renewable and Sustainable Energy Reviews. 2015;**42**:1362-1369. DOI: 10.1016/j.rser.2014.11.009

[110] International Energy Agency [Internet]. Available from: https://www.iea.org/energyaccess/database/ [Accessed: June 28, 2018]

[111] Najjar YSH, Kseibi MM. Thermoelectric stoves for poor deprived regions – A review. Renewable and Sustainable Energy Reviews. 2017;**80**: 597-602. DOI: 10.1016/j.rser.2017.05.211

[112] Gao HB, Huang GH, Li HJ, Qu ZG, Zhang YZ. Development of stove-powered thermoelectric generators: A review. Applied Thermal Engineering. 2016;**96**:297-310. DOI: 10.1016/j.applthermaleng.2015.11.032

[113] Deasy MJ, O'Shaughnessy SM, Archer L, Robinson AJ. Electricity generation from a biomass cook stove with MPPT power management and passive liquid cooling. Energy for Sustainable Development. 2018;**43**: 162-172. DOI: 10.1016/j.esd.2018.01.004

[114] Date A, Date A, Dixon C, Akbarzadeh A. Progress of

thermoelectric power generation systems: Prospect for small to medium scale power generation. Renewable and Sustainable Energy Reviews. 2014;**33**: 371-381. DOI: 10.1016/j.rser.2014.01.081

[115] Kraemer D, McEnaney K, Chiesa M, Chen G. Modeling and optimization of solar thermoelectric generators for terrestrial applications. Solar Energy. 2012;**86**(5):1338-1350. DOI: 10.1016/j.solener.2012.01.025

[116] Olsen ML, Warren EL, Parilla PA, Toberer ES, Kennedy CE, Snyder GJ, et al. A high-temperature, high-efficiency solar thermoelectric generator prototype. Energy Procedia. 2014;**49**:1460-1469. DOI: 10.1016/j.egypro.2014.03.155

[117] Baranowski LL, Snyder GJ, Toberer ES. Concentrated solar thermoelectric generators. Energy & Environmental Science. 2012;**5**:9055-9067. DOI: 10.1039/C2EE22248E

[118] Molina MG, Juanicó LE, Rinalde GF. Design of innovative power conditioning system for the grid integration of thermoelectric generators. International Journal of Hydrogen Energy. 2012;**37**(13): 10057-10063. DOI: 10.1016/j.ijhydene.2012.01.177

Chapter 2

# Definition and Design of Zero Energy Buildings

*Yuehong Lu, Xiao-Ping Zhang, Zhijia Huang, Jinli Lu and Changlong Wang*

## Abstract

The wide application of renewable energy system (RES) in buildings combined with numerous financial incentives on RES paves the way for future zero energy buildings (ZEB). Although the definition of ZEB still lacks a national building code and international standards, the number of ZEB projects is still increasing worldwide which seems to be the pioneer ZEB buildings. However, due to the intermittency of the renewable resources, various uncertain parameters, and dynamic electricity price from the grid, how to select the renewable energy system for buildings is one of the challenges and therefore becomes an extensive concern for both researchers and designers. In addition, questions like how to achieve the target of zero energy for different types of buildings, should the building be designed as an independent ZEB or a group of buildings to be a ZEB cluster, and how to make building owners actively involved in installing enough RES for the building are still on the air. This chapter will present a comprehensive view on several key issues related with ZEB, that is, definition, evaluation criteria, design method, and uncertainty analysis, and the penalty cost scheme is also proposed for consideration as one policy to assist the promotion of ZEB.

**Keywords:** renewable energy, optimal design, zero energy building, robust method, feed-in tariff, penalty cost

## 1. Introduction

Building energy consumption is generally recognized as one of the main sectors contributing to the whole primary energy consumption and greenhouse gas emissions in the world, which greatly raise public awareness on building energy conservation in recent years. Green building (GB), low-energy building (LEB), and nearly/net-zero energy building (ZEB/nZEB/NZEB) are widely developed for the advantages of low-energy demand in the building, efficient energy system operation, and integration of renewable energy system [1–5]. In addition, numerous incentive policies, such as investment subsidies, feed-in tariff, net-metering schemes, etc., have been applied to promote the application of renewable energy sources [6–16], thus paving the way for future zero target for buildings.

The concept of ZEB/nZEB is extended from autonomous buildings that are targeted to operate off-grid by installing enough solar PV and/or wind turbine for the generation of all the energy the building required to include grid-connected ZEB that is aimed to balance annual energy exchange with the grid. The off-grid ZEB has also been named "autonomous" or "stand-alone" building as shown in

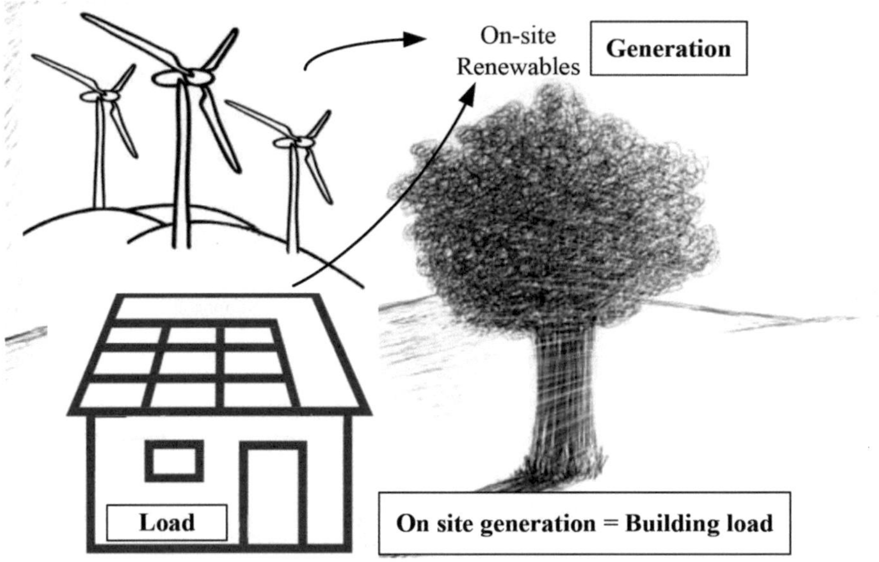


**Figure 1.**
*Basic elements in the definition of off-grid zero energy building.*

**Figure 1**, which can be defined as "Zero Stand Alone Buildings are buildings that do not require connection to the grid or only as a backup. Stand-alone buildings can autonomously supply themselves with energy, as they have the capacity to store energy for night-time or wintertime use" [17].

The on-grid ZEB is a "grid-connected" or "grid-integrated" net-zero energy building that is connected to one or more energy infrastructure as shown in **Figure 2**; it can be defined as "Zero Net Energy Buildings are buildings that over a year are neutral, meaning that they deliver as much energy to the supply grids as they use from the grids. Seen in these terms they do not need any fossil fuel for heating, cooling, lighting or other energy uses although they sometimes draw energy from the grid" [17].

However, no national ZEB codes and international standards have been developed since numerous proposed approaches spotlight different aspects of ZEB. The metric applied for the "zero" balance is a vital issue since it affects how renewable energy system will be selected to achieve this goal. Torcellini et al. [18] introduced four different nZEB definitions, including site nZEB, source nZEB, emissions nZEB, and cost nZEB, as defined below:

- Site nZEB: A site nZEB produces at least as much energy as it uses in a year when accounted for at the site.

- Source nZEB: A source nZEB produces at least as much energy as it uses in a year when accounted for at the source. Source energy refers to the primary energy used to generate and deliver the energy to the site.

- Emission nZEB: A net-zero emission building produces at least as much emission-free renewable energy as it uses from emission-producing energy sources.

- Cost nZEB: In a cost nZEB, the amount of money the utility pays the building owner for the energy the building exports to the grid is at least equal to the amount the owner pays the utility for the energy services and energy used over the year.

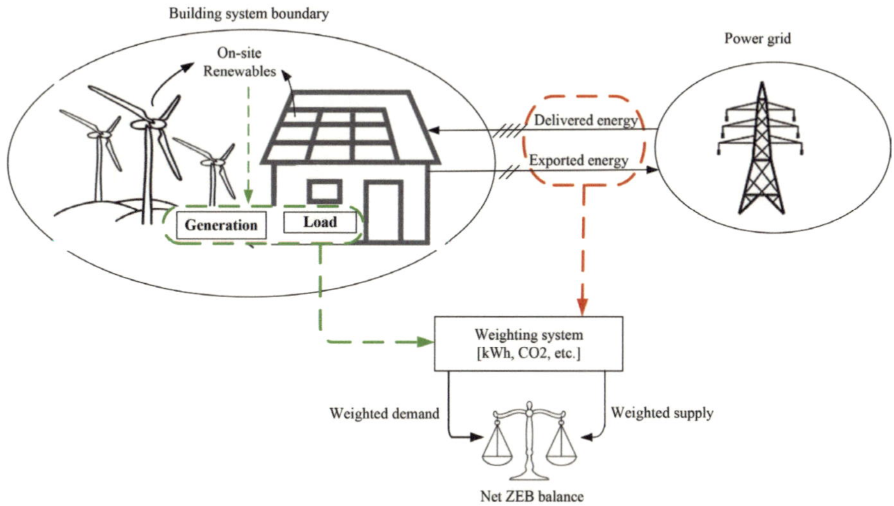


**Figure 2.**
*Basic elements in the definition of on-grid net-zero energy building.*

The concept of balance can be defined in the following mathematical equations, the balance between export and import energies (Eq. (1)) or the balance between load and generation (Eq. (2)) [19]. The balance between load and generation is generally used as a basic requirement during the design phase of ZEB. By contrast, the balance between export and import energies is particularly useful for evaluating its matchability between load and generation and the interaction between building and grid. In the previous study, most studies defined "net" as the building energy consumption and RES generation to be equal:

$$\text{Net energy} = /\text{Export}/-/\text{Import}/ \geq 0 \tag{1}$$

$$\text{i.e.} \sum_i \text{Exported\_energy}(i) \times \text{weight}(i) - \sum_i \text{Imported\_energy}(i) \times \text{weight}(i) \geq 0$$

$$\text{Net energy} = /\text{Generation}/-/\text{Load}/ \geq 0 \tag{2}$$

$$\text{i.e.} \sum_i \text{Generation\_energy}(i) \times \text{weight}(i) - \sum_i \text{Load\_energy}(i) \times \text{weight}(i) \geq 0$$

## 2. Incentive policies

Substantial policies and regulations have been provided to support the installation of RES power plants and thus simulate the widespread of ZEB applications. Under the support of these policies, an increasing number of ZEB case studies have been conducted, and there are over 360 ZEB projects which had been identified in different countries till 2013, as shown in **Figure 3**.

The continued growth in ZEB projects is mainly driven by the progressive financial incentives on renewable energy promotion, which is summarized for several countries from different parts of the world, as shown in **Table 1**. The main support policies on RES in different countries are described as follows [6]:

- Investment subsidies: Based on a percentage of the renewable energy output or the specific investment upfront cost.

**Figure 3.**
*World map of more than 360 internationally known net-zero energy buildings [20].*

| Incentive policies | Australia | Belgium | China | France | Germany | Italy | Japan | Spain | The United Kingdom | The United States |
|---|---|---|---|---|---|---|---|---|---|---|
| Renewable energy targets | O | O | R | R | O | O | R | O | O | R* |
| Feed-in tariff/ premium payment | • | | R | R | R | R | R | | O | R* |
| Electric utility quota obligation/ RPS | O | • | O | | | | O | | O | R* |
| Net metering | | • | | | | O | O | O | | R* |
| Tradable REC | O | O | | O | | O | O | O | O | • |
| Capital subsidy, grant, or rebate | O | • | O | O | O | O | O | O | O | O |
| Reductions in sales, energy, $CO_2$, VAT, or other taxes | | O | O | O | O | O | | | O | O |
| Public investment, loans, or grants | O | | O | O | O | O | O | | O | O |

*O, existing national (may also include state/provincial); •, existing subregional (e.g., state/provincial); R, revised (*indicates state/provincial).*

**Table 1.**
*Promotion policies of renewable energy in several countries [21].*

- Feed-in tariff: The producer receives total payments per kWh of generated electricity at a fixed price. It is guaranteed by the government.

- Net-metering schemes: Net-metering (NM) and self-consumption (SC) schemes. Billing agreement between utilities and their customers to feed electricity the producer does not use back into the grid.

- Tradable Green Certificates: Certificates that can be sold in the market, allowing RES generators to obtain revenue, in addition to the earnings from the sale of electricity fed into the grid.

## 3. Design methodologies

### 3.1 Design step

Although no exact approach has been developed for the target of zero balance during the design phase of ZEB, there are still some consensus and several common design elements for designing ZEB. A thorough design approach was proposed which involves 12 steps containing four foundational procedures, that is, applied metrics (e.g., primary energy, the cost), passive design (e.g., building envelope, orientation), active design (e.g., HVAC, lighting), and renewable energy system design (e.g., photovoltaic panel, wind turbine) for the design of ZEB [22, 23], as shown in **Figure 4**. Theoretically, design optimization of ZEB should be conducted considering the three vital design steps, that is, steps 7, 8, and 9, simultaneously to obtain a comprehensive combined design option for ZEB. Therefore, design optimization for ZEB is usually solved by integration of two or more software.

Passive design is an important procedure to reduce the building energy demand as much as possible. Then, high-efficiency active energy systems such as heating, cooling, and ventilation systems and lighting systems should be applied and improved together with high-performance control strategies; these could further reduce operational energy consumption in the building. Lastly, the feasibility of renewable energy technology should be assessed and selected as an on-site power supply system which works together with the power grid to reach the target of zero energy demand.

Various software tools have been developed to facilitate the selection of passive design, active design, and RES for buildings; several popular software are listed and compared in **Tables 2** and **3**. In ZEB design, the building energy demand can be firstly evaluated by using building energy simulation software such as EnergyPlus or TRNSYS. The selection of suitable renewable energy system for the building can then be conducted in software such as HOMER and TRNSYS. The design optimization software, HOMER, is developed by the US National Renewable Energy Laboratory (NREL) to assist in design optimization of hybrid energy systems for both grid-connected and autonomous building based on net present cost [24–27]. However, HOMER can only address a single-objective function for minimizing the net present cost, and it cannot solve multi-objective problems [28].

### 3.2 Performance evaluation criteria

It is important to determine the evaluation criteria at the design stage. Various criteria have been proposed from a different perspective of users, which can be classified into four aspects covering technical and environmental issues, economic factors, and the interaction between building and grid, as shown in **Figure 5**.

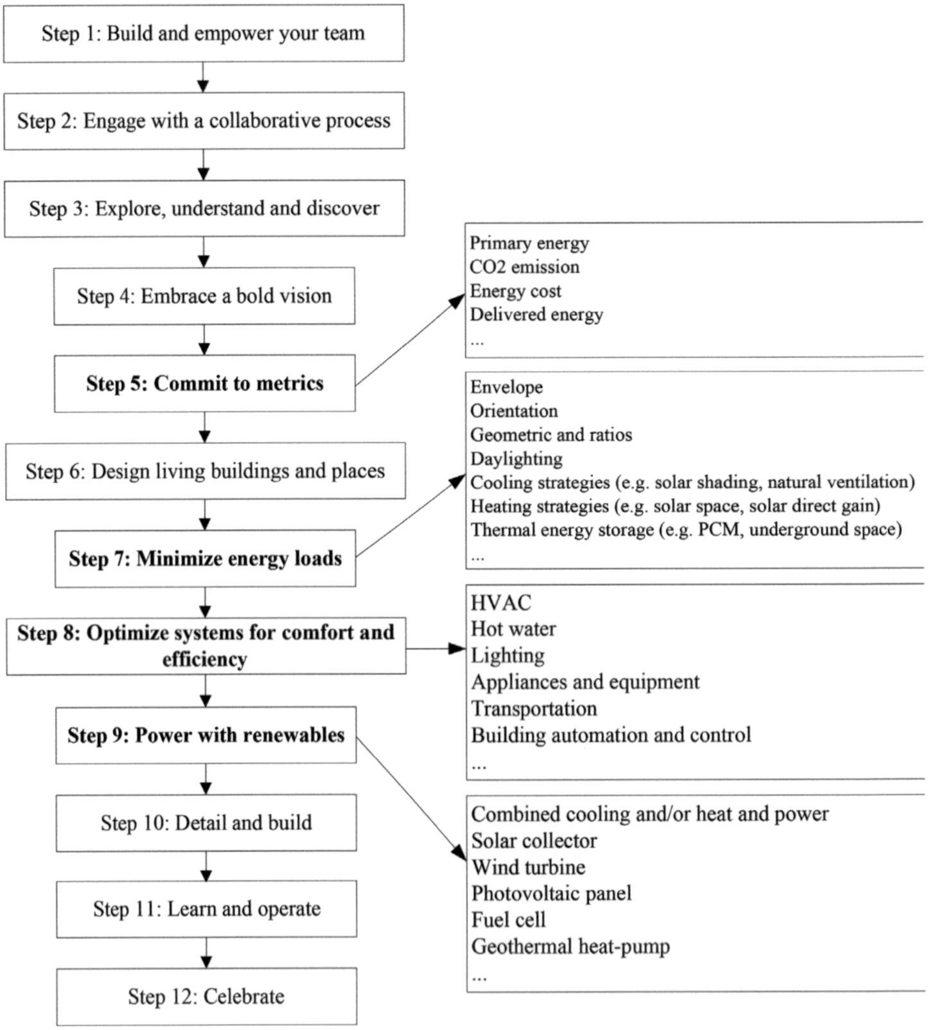


**Figure 4.**
*The main steps to designing nZEB.*

| | DOE-2 | eQUEST | EnergyPlus | ESP-r | DeST | TRNSYS |
|---|---|---|---|---|---|---|
| Room heat balance calculation | | √ | √ | √ | √ | √ |
| Humidity calculation | | √ | √ | √ | √ | √ |
| Heat comfort calculation | | | √ | √ | √ | √ |
| Nature ventilation calculation | | √ | √ | √ | √ | √ |
| Sunlight analysis | √ | √ | √ | √ | √ | √ |
| Greenhouse gas | √ | √ | √ | √ | √ | √ |
| Connection with other software | | | √ | √ | √ | √ |

**Table 2.**
*Comparison of building energy consumption software [29].*

| | HOMER | HYBRID2 | iHOGA | RETScreen |
|---|---|---|---|---|
| Economical analysis | √ | √ | √ | √ |
| Technical analysis | √ | √ | √ | √ |
| PV system | √ | √ | √ | √ |
| WT system | √ | √ | √ | √ |
| Generator set | √ | √ | √ | |
| Storage device | √ | √ | √ | √ |
| Bioenergy | √ | | | |
| Hydro energy | √ | | √ | |
| Thermal system | | √ | | |
| Advantage | User-friendly, easy to understand, efficient graphical representation of results, hourly data-handling capacity | User-friendly, multiple electrical load options, detailed dispatching option | multi- or mono-objective optimization, sensitivity analysis, low computational time, net balance | User-friendly, strong product database, financial analysis |

**Table 3.**
*Comparison of renewable energy simulation software [30].*

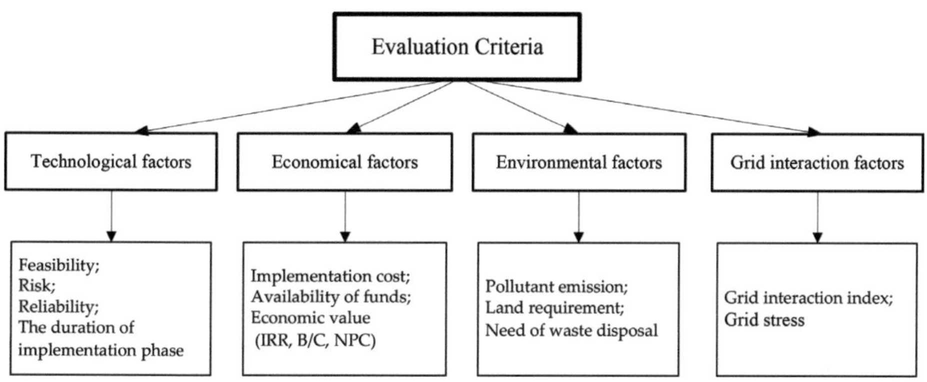


**Figure 5.**
*The main four factors for evaluating ZEB performance.*

In terms of technological factors, recent researches have been focused on feasibility and/or reliability study of different technologies (passive design, active design, RES) for ZEB.

- Feasibility: Available technologies should be assessed to identify the possibility for the building to achieve zero energy building for a particular region.

- Reliability: The criterion estimates the ability of the selected combined technologies for the building to perform its required function for a specified time.

In terms of economic factors, ZEB users are more concerned about the economic value, especially the cost and its payback period of installing on-site RES.

- Economic value [life cycle analysis (LCA), net present cost (NPC)]: The proposed renewable energy alternative will be assessed using one of the engineering economic techniques which are net present cost (NPC), life cycle analysis (LCA), benefit/cost analysis, and payback period.

In terms of environmental factors, the reduction of building load will definitely reduce the energy required from the grid and on-site RES size, which can be measured as pollutant emission.

- Pollutant emission: The criterion measures the equivalent emission of $CO_2$, air emissions which are the results of applying different technologies in ZEB for a particular period.

In terms of grid interaction factors, the two-way electricity flow between building and grid poses more than technological challenges; those ZEB homeowners may make heavier use of the grid than the conventional building under one-way power flow. Grid interaction index is one of the indicators used to assess the grid stress caused by ZEB.

- Grid interaction index (GII): The criterion is defined as the standard deviation of the building-grid interaction over the specified time (e.g., 1 year). It is used to estimate the average stress of building on the grid, and a low standard deviation is preferred [28].

### 3.3 Optimization method

#### *3.3.1 Single-objective design optimization*

It is reported that there are more than 50% of design optimization problems that are addressed as single-objective optimization problems, and they are usually focused on the most important criteria such as economic cost or environmental issues. For designing ZEB, the optimization may be conducted by focusing on the only one aspect of ZEB performance, e.g., NPC and $CO_2$ emissions. Besides, since multi-objective design optimization problems can also be transformed into single-objective optimization problems by using weighting factor, it is reasonable to convert all of the concerned ZEB performance indices into one combined function, as shown in (Eq. (3)). Where X represents a vector of design variables at the design stage, $f_{ave}$ and $f_i$ ($i = 1, 2 \ldots n$) are the combined objective and the normalized sub-objectives, respectively; $w_i$ is the corresponding weighting factor for each sub-objective:

$$\text{Min} f_{ave} = w_1 \times f_1(X) + w_2 \times f_2(X) + \ldots + w_n \times f_n(X) \tag{3}$$

$$\text{s.t.} AX \le a \tag{4}$$

$$g_1(X) \ge 0 \tag{5}$$

$$g_2(X) = 0 \tag{6}$$

#### *3.3.2 Multi-objective design optimization*

The design and operation of ZEB are actually integrated with building owners, environment, energy source, and smart grid; it is, therefore, a multi-objective design optimization problem with even contradicting objectives. In general, genetic algorithm (GA) is the most popular optimization approach for single-objective and multi-objective optimizations of energy systems in numerous studies [31, 32]. Besides, particle swarm optimization (PSO) is another favored method for optimal design of energy systems in recent papers [33, 34]. A typical optimization process is shown in **Figure 6**.

### 3.4 Penalty cost for ZEB

Although the progressive incentive policies have been recognized to widely encourage the installation of renewable energy system for buildings, the financial support scheme is forecasted to be downtrend and RES cost to be high, which are a barrier for promoting future buildings to be zero energy buildings. Therefore, a penalty cost scheme may be a good solution to build up the public's confidence and encourage them to be actively involved in ZEB application.

A comparison of the building cost under different mismatch ratios is shown in **Figure 7**. It is found that the minimum total cost is supposed to be located in O1 under mismatch ratio less than 0, possibly between −1.0 and 0.0, indicating that the selection of design option under mismatch ratio of 1.0 is not cost-effective. However, the introduction of penalty cost can move the minimum cost from O1 to O2, or the higher mismatch ratio the less cost, depending on the type of penalty cost designed by designers.

The total cost (*TC*) of the building basically consists of the initial cost (*IC*) of RES (e.g., PV, WT) and the operation cost (*OC*) during the building usage stage due to the electricity consumption from grid and oil consumption (Eq. (4)). The penalty cost can be expressed as a mathematic expression, which is assumed to follow a segmented function, as shown in Eq. (5). It should be noted that the cost results may be greatly different according to the designed penalty cost:

$$TC_P = IC + OC + PC \tag{7}$$

$$PC = \begin{cases} TC_{1.0} \times (a - \alpha \times SF), SF < p_1 \\ TC_{1.0} \times (b - \beta \times SF), p_1 \le SF < p_2 \\ TC_{1.0} \times (c - \delta \times SF), SF \ge p_2 \end{cases}. \tag{8}$$

### 3.5 Individual ZEB or ZEB cluster

In recent years, wide ranges of researches are available on the topic of RES design/control for an off-grid building or a village in a remote area [25–27].

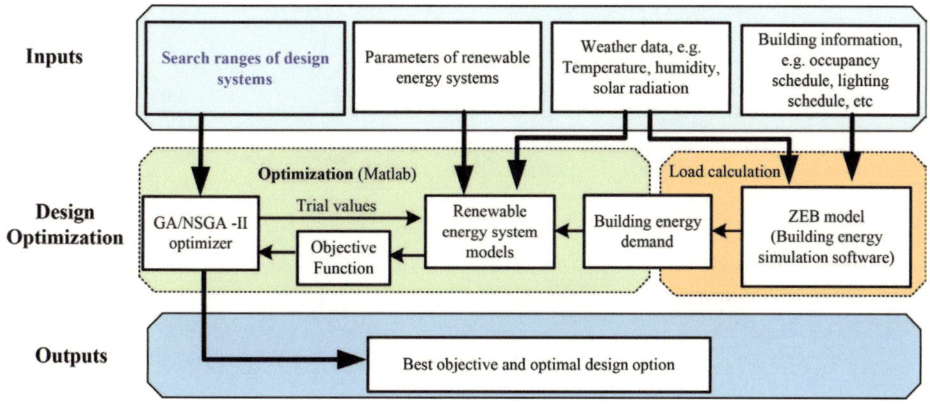


**Figure 6.**
*Single-/multi-objective design optimization using GA/NSGA-II.*

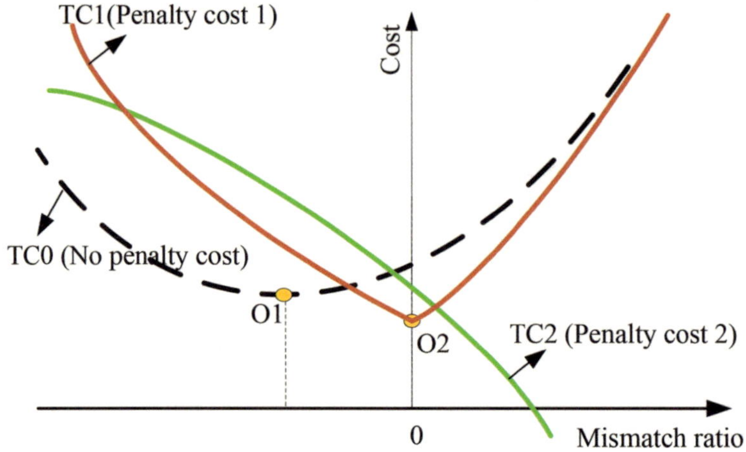


**Figure 7.**
*The cost of building under different mismatch ratios.*

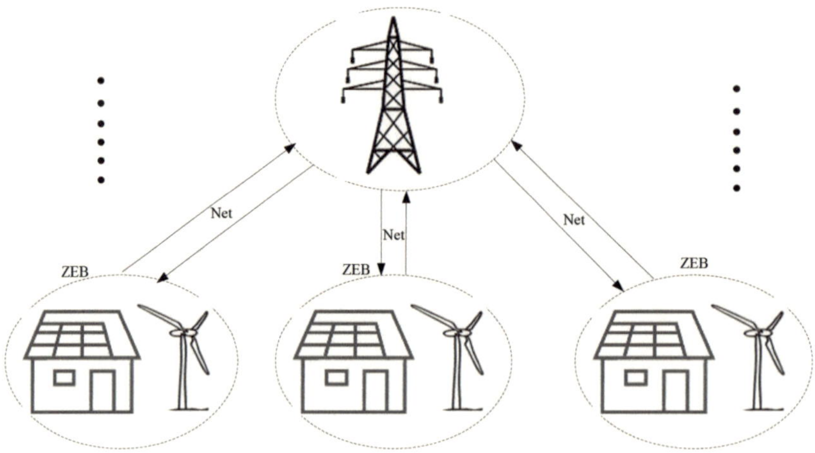


**Figure 8.**
*A typical diagram of individual on-grid ZEB.*

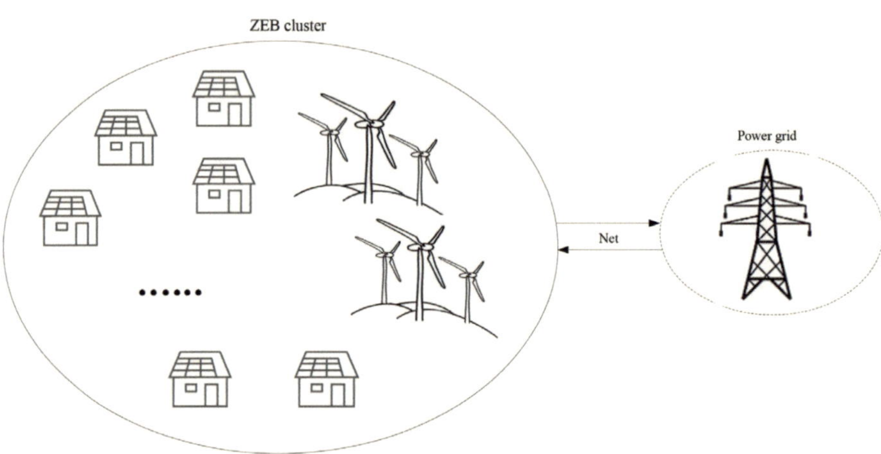


**Figure 9.**
*A typical diagram of on-grid ZEB cluster.*

However, the question is, are the buildings better to be designed as individual ZEB separately (**Figure 8**) or a ZEB cluster (**Figure 9**) when the grid power is available? By considering the dynamic electricity price from the power grid and dynamic financial incentives of sell back price from RES, the investment of RES is supposed to be different, while electricity exchange between the building and the grid is also supposed to be greatly different since the building in ZEB cluster can share on-site generation among these grouped buildings. In the study of Sun et al. [35], performance potentials are investigated by comparing single-building level using non-collaborative controls and building-group level using collaborative controls. However, a systematic and comprehensive comparison of the differences between design/control strategies for individual on-grid ZEB and on-grid ZEB cluster should be further explored and formed for ZEB development.

## 4. Uncertainty analyses

### 4.1 Uncertain parameters

ZEB is generalized as a type of sensitive building since its target is affected not only by the variation of building energy load but also by the fluctuation of local renewable energy resources. In general, a comprehensive sensitivity analysis is required to be conducted by considering both on-site generation and building energy load, as shown in **Figure 10**.

In terms of on-site RES generation, the availability of renewable resources (e.g., solar radiation and/or wind speed) and RES parameters (e.g., PV/WT efficiency and life time) can directly affect on-site energy generation for building electricity supply and thus greatly affect RES selection as well as the corresponding building performance.

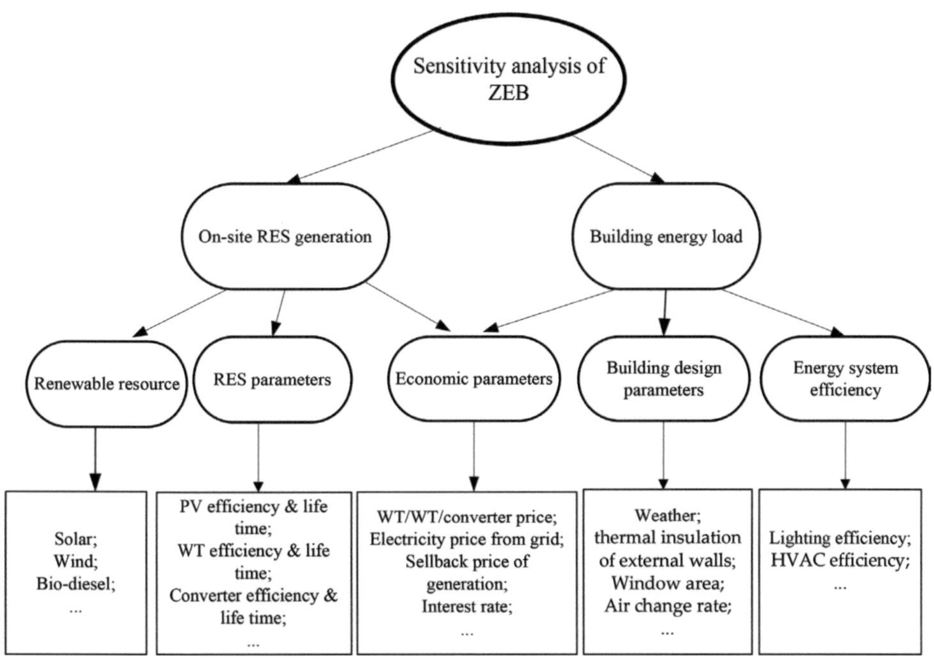


**Figure 10.**
*Four aspects in terms of sensitivity analysis for ZEB.*

In terms of building energy load, building design parameters (e.g., indoor set temperature and humidity, thermal insulation of external walls, window area, etc.), operational parameters (e.g., outdoor temperature and humidity, solar radiation, etc.), and energy system efficiency (e.g., lighting efficiency, HVAC efficiency, etc.) are the main parameters affecting building load.

Economic parameters including the price of RES, feed-in tariff, and electricity price from/to the grid are also key parameters affecting RES selection and ZEB performance. In addition, many researches have found that both the building energy load and the local renewable resources are different even for the same building located in different climate zones, which indicate that different key design parameters may exist for different climate zones and should be further identified.

### 4.2 Robust design

Therefore, there are many uncertain parameters which may cause great performance discrepancy between the design stage and operation stage of ZEB. The impact of uncertain parameters on system selection can be illustrated in **Figure 11**. In convention building, since it has no constraint on the mismatch between building energy consumption and on-site generation, the optimal design option O is usually selected within all the design options (Area of A), which is located below the net-zero balance line with 100% confidence level. When designing ZEB using deterministic approach, the constraint of annual energy balance is achieved for the design year, and thus the optimal design option O′ is usually selected within a few design options (Area of B), which is located on the net-zero balance line with approximately 50% confidence level. When uncertain parameters are concerned for a robust ZEB design, a narrowed area of C is identified as the selected option is required to satisfy many years of operation. Therefore, the optimal design option O″ is selected on/above the net-zero balance line with 100% confidence level. In the study of Lu [36], the RES size should be selected with a mismatch ratio of about 30% to ensure a probability of 100% of being ZEB in different years.

### 4.3 Impact of key parameters on ZEB performance

As mentioned in **Figure 12**, various parameters can affect ZEB system selection and performance. Six key parameters are selected here and grouped in pairs according to their specifications, i.e., PV price (ranges between 1200 and 2000 $/kW) and

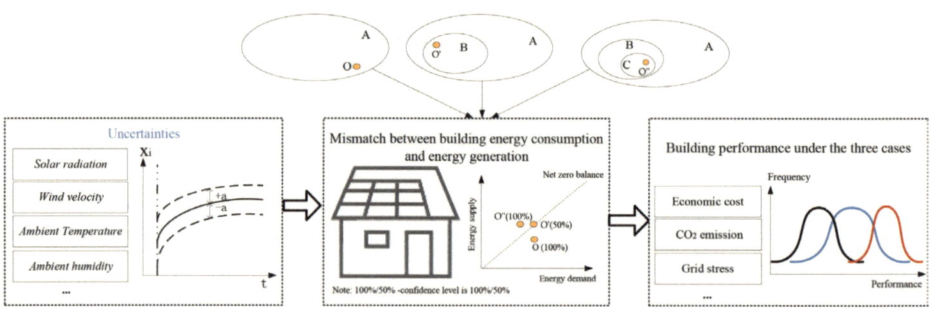


**Figure 11.**
*Impact of uncertain parameters on ZEB performance. Note: A, all design options; B, design options for nZEB based on deterministic condition; C, design options for nZEB under uncertainties; O/O′/O″, optimal design option.*

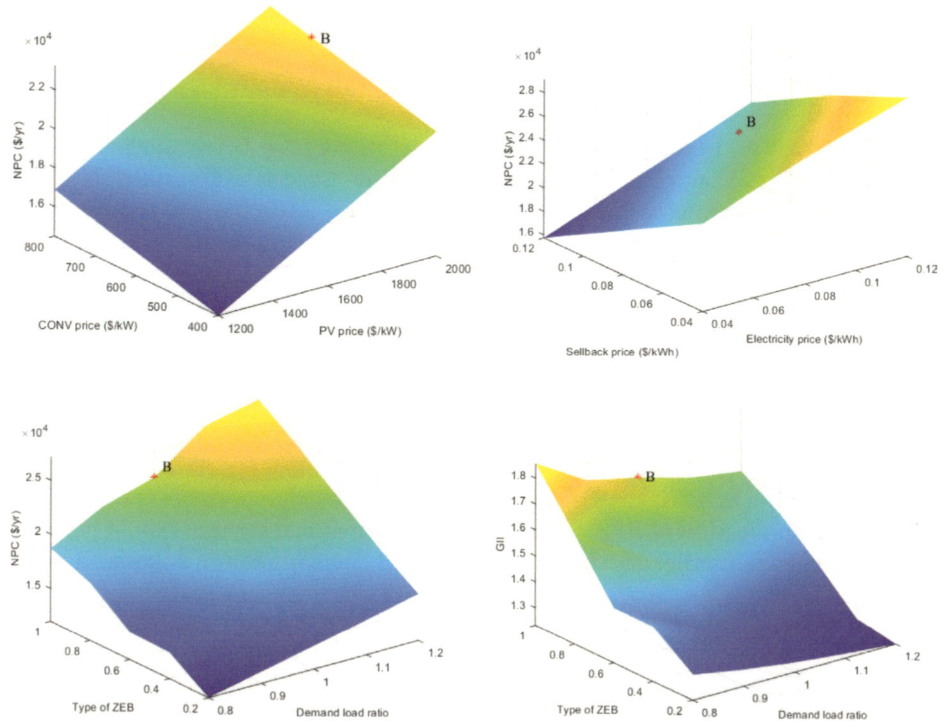


**Figure 12.**
*Variations of NPC and GII under the effect of two factors (Note: The point B is the value under basic case.).*

converter price (ranges between 400 and 800 $/kW), electricity price (EP) (ranges between 0.04 and 0.12 $/kWh) and sellback price (SP) (ranges between 0.04 and 0.12 $/kWh), and demand load ratio (ranges between 0.8 and 1.2) and the type of ZEB (ranges between 0.2 and 1.0), respectively. The impact of the group in pairs on NPC and GII is compared based on a hypothetical residential house (an area of 120 $m^2$) that is located in Shanghai, China. Under the selected ranges of parameters, electricity price, sellback price, demand load ratio and the type of ZEB are identified to be more important on NPC than PV and CONV price. It should be noted that the results may be different when applied for different parameter variation ranges.

## 5. Conclusion

This chapter aims to present a comprehensive view of the key issues related to the development of zero energy building including the definition, supporting incentives, evaluation criteria, design methodologies, and uncertainty analysis. Although a wide range of researches can be found to investigate one aspect of the ZEB study, there are still a lot of challenges faced to be solved in the future:

1. A consensus definition and interpretation of national/regional ZEB should be further declared; then, the design/control strategies with the corresponding performance of ZEB can be investigated and compared under the same standard.

2. Since a lot of factors/parameters can cause the discrepancies between predicted and realized target and ZEB performance, it should be noted that even the same factors may have a different effect on ZEB design for a specified

region considering the applied energy systems and its financial support. So, it is necessary to identify and classify the key factors affecting ZEB performance for different conditions.

3. The future grid-connected ZEB seems to definitely pose a great challenge for smart grid considering the new complex energy usage as well as on-site RES generation features of ZEB. The stress caused by ZEB on the grid is different from conventional buildings, which should be further identified and taken into consideration on ZEB design.

4. ZEB is generally catering for an individual building, while the investigation of zero energy building cluster, zero energy community, and zero energy high-rise building is still required and thus forms some standards for the design/control strategies in each type of ZEB application.

5. The existing financial incentives are mostly proposed to promote the wide-spread application of RES, while a systematic financial scheme should be developed to further assist and stimulate ZEB development in a standard and rapid way.

## Acknowledgements

Our thanks to the National Natural Science Foundation of China (Project No. 51608001 and Project No. 51478001) for financial support on this research work reported in this chapter. The authors also acknowledge the support from the China Scholarship Council (CSC) and the Anhui University of Technology and research support from the University of Birmingham in the UK.

## Author details

Yuehong Lu[1,2]*, Xiao-Ping Zhang[2], Zhijia Huang[1], Jinli Lu[1] and Changlong Wang[1]

1 Department of Civil Engineering and Architecture, Anhui University of Technology, Ma'anshan, China

2 Department of Electronic, Electrical and Systems Engineering, University of Birmingham, Birmingham, UK

*Address all correspondence to: luyuehongtuzi@163.com

## References


[1] Li XW, Wen J. Net-zero energy impact building clusters emulator for operation strategy development. In: 2014 ASHRAE Annual Conference, at Seattle, WA, USA; June 28–July 02; 2014

[2] Sun YJ, Huang P, Huang GS. A multi-criteria system design optimization for net zero energy buildings under uncertainties. Energy and Buildings. 2015;**97**:196-204

[3] Aelenei L, Gonçalves H. From solar building design to net zero energy buildings: Performance insights of an office building. Energy Procedia. 2014;**48**:1236-1243

[4] Wang W, Zmeureanu RG, Rivard H. Applying multi-objective genetic algorithms in green building design optimization. Building and Environment. 2005;**40**:1512-1525

[5] Zebra. Nearly Zero-energy Building Strategy 2020. September 26, 2014. http://zebra2020.eu/

[6] Javier Ramírez F, Honrubia-Escribano A, Gómez-Lázaro E, Pham Duc T. Combining feed-in tariffs and net-metering schemes to balance development in adoption of photovoltaic energy: Comparative economic assessment and policy implications for European countries. Energy Policy. 2017;**102**:440-452

[7] Bakhshi R, Sadeh J. Economic evaluation of grid-connected photovoltaic systems viability under a new dynamic feed-in tariff scheme: A case study in Iran. Renewable Energy. 2018;**119**:354-364

[8] Promoting Renewable Energy Sources in EU after 2020. Briefing EU Legislation in Progress. 2018. http://www.europarl.europa.eu/RegData/etudes/BRIE/2017/599278/EPRS_BRI(2017)599278_EN.pdf

[9] Zhang MM, Zhou DQ, Zhou P, Liu GQ. Optimal feed-in tariff for solar photovoltaic power generation in China: A real options analysis. Energy Policy. 2016;**97**:81-192

[10] Lau KY, Muhamad NA, Arief YZ, Tan CW, Yatim AHM. Grid-connected photovoltaic systems for Malaysian residential sector: Effects of component costs, feed-in tariffs, and carbon taxes. Energy. 2016;**102**:65-82

[11] UK introduces feed-in tariffs. https://www.ofgem.gov.uk/environmental-programmes/fit/fit-tariff-rates

[12] Renewables 2014 Global renewable status report. REN21 (Renewable Energy Policy Network for the 21st Century) [Internet]. 2014. Available from: http://www.ren21.net/Portals/0/documents/Resources/GSR/2014/GSR2014_full%20report_low%20res.pdf

[13] Lüthi S, Wüstenhagen R. The price of policy risk–Empirical insights from choice experiments with European photovoltaic project developers. Energy Economics. 2012;**34**:1001-1011

[14] Huang YH, Wu JH. Assessment of the feed-in tariff mechanism for renewable energies in Taiwan. Energy Policy. 2012;**39**:8106-8115

[15] Pyrgou A, Kylili A, Fokaides PA. The future of the Feed-in Tariff (FiT) scheme in Europe: The case of photovoltaics. Energy Policy. 2016;**95**:94-102

[16] Abolhosseini S, Heshmati A. The main support mechanisms to finance renewable energy development. Renewable and Sustainable Energy Reviews. 2014;**40**:876-885

[17] Laustsen J. Energy efficiency requirements in building codes. In:

Energy Efficiency Policies for New Buildings, OECD. Paris: IEA; 2008

[18] Torcellini P, Pless S, Deru M, Crawley D. Zero Energy Buildings: A Critical Look at the Definition. ACEEE Summer Study, Pacific Grove, California; 2006. pp. 14-18

[19] Deng S, Wang RZ, Dai YJ. How to evaluate performance of net zero energy building–A literature research. Energy. 2014;**71**:1-16

[20] World Map of nZEBs [Internet]. 2013. Available from: http://batchgeo.com/map/net-zero-energy-buildings

[21] Taxes and Incentives for Renewable Energy [Internet]. 2015. Available from: https://home.kpmg.com/xx/en/home/services/tax/energy-tax.html

[22] Rodriguez-Ubinas E, Montero C, Porteros M, Vega S, Navarro I, Castillo-Cagigal M, et al. Passive design strategies and performance of Net Energy Plus Houses. Energy and Buildings. 2014;**83**:10-22

[23] Doust N, Masera G, Frontini F, Imperadori M. Cost optimization of a nearly net zero energy building: A case study. In: SIMUL 2012, the Fourth International Conference on Advances in System Simulation. 2012. pp. 44-49

[24] Iqbal MT. A feasibility study of a zero energy home in Newfoundland. Renewable Energy. 2004;**29**(2):277-289

[25] Hassoun A, Dincer I. Development of power system designs for a net zero energy house. Energy and Buildings. 2014;**73**:120-129

[26] Li C, Ge XF, Zheng Y, Xu C, Ren Y, Song CG, et al. Techno-economic feasibility study of autonomous hybrid wind/PV/battery power system for a household in Urumqi, China. Energy. 2013;**55**:263-272

[27] Rezzouk H, Mellit A. Feasibility study and sensitivity analysis of a stand-alone photovoltaic–diesel–battery hybrid energy system in the north of Algeria. Renewable and Sustainable Energy Reviews. 2015;**43**:1134-1150

[28] Lu YH, Wang SW, Zhao Y, Yan CC. Renewable energy system optimization of low/zero energy buildings using single-objective and multi-objective optimization methods. Energy and Buildings. 2015;**89**:61-75

[29] Han Y, Liu X, Chang L. Comparison of software for building energy simulation. Journal of Chemical and Pharmaceutical Research. 2014;**6**(3):467-471

[30] Sinha S, Chandel SS. Review of software tools for hybrid renewable energysystems. Renewable and Sustainable Energy Reviews. 2014;**32**:192-205

[31] Palonen M, Hasan A, Siren K. A genetic algorithm for optimization of build-ing envelope and HVAC system parameters. In: IBPSA Conference on Eighth International Building Performance Simulation Association. 2009. pp.159-166

[32] Deb K, Pratap A, Agarwal S, Meyarivan T. A fast and elitist multiobjectivegenetic algorithm: NSGA-II. IEEE Transactions on Evolutionary Computation. 2002;**6**(2):182-197

[33] Avril S, Arnaud G, Florentin A, Vinard M. Multi-objective optimization of batteries and hydrogen storage technologies for remote photovoltaic systems. Energy. 2010;**35**:5300-5308

[34] Moghaddas-Tafreshi SM, Hakimi SM. Optimal sizing of a stand-alone hybrid power system via particle swarm optimization for Kahnouj area

in south-east of Iran, 2009. Renewable Energy. 2009;**34**:1855-1862

[35] Sun Y, Huang G, Xu X, Lai AC. Building-group-level performance evaluations of net zero energy buildings with non-collaborative controls. Applied Energy. 2018;**212**:565-576

[36] Lu YH, Wang SW, Yan CC, Huang ZJ. Robust optimal design of renewable energy system in nearly/net zero energy buildings under uncertainties. Applied Energy. 2017;**187**:62-71

Chapter 3

# Energy and Seismic Rehabilitation of RC Buildings through an Integrated Approach: An Application Case Study

*Antonio D'Angola, Vincenzo Manfredi, Angelo Masi and Marianna Mecca*

## Abstract

The high number of existing buildings in Italy without adequate seismic and thermal performances requires the definition of integrated retrofitting techniques in order to improve the seismic performance and to reduce energy losses at the same time. On one hand, an integrated approach appears mandatory considering that improving only the energy efficiency of nonseismic buildings leads to an increase of their exposure and, therefore, of their risk in the case of seismic events. On the other hand, seismic strengthening without an adequate thermal assessment and rehabilitation could compromise living comfort and energy maintenance costs. In this context, an application of integrated approach for the rehabilitation of reinforced concrete (RC) existing buildings has been proposed referring to a case study representative of the Italian building stock. Different configurations of infill panels have been considered in order to analyze both energy and seismic performance. Monthly quasi-steady state and hourly dynamic models have been used for the calculation of the energy need of buildings located in different Italian climate and seismic zones. Seismic performances have been evaluated by means of incremental nonlinear dynamic analysis (IDA). As-built and post-retrofit performances have been compared in order to evaluate the effectiveness of the proposed intervention solutions.

**Keywords:** existing RC buildings, integrated approach, thermal insulation, energy efficiency, seismic strengthening, infill walls

## 1. Introduction

In Europe, about 25 billion m$^2$ of available floor space (data relevant to 27 EU member states in 2011 plus Switzerland and Norway) was built up in the past. A predominant part of this surface, about 75%, is made up of residential buildings (64% single family houses, 36% apartment blocks), while the other part (25%) consists of more complex and heterogeneous nonresidential buildings. More than 40% of residential buildings were constructed before the 1960s, with the largest percentages of older buildings in the United Kingdom, Denmark, Sweden, France,



IntechOpen

Czech Republic, and Bulgaria [1]. As a result, beyond the well-known deficit of seismic protection, dramatically pointed out by past earthquakes, also the energy performances of the European buildings are expected to be generally poor. In fact, energy consumed in buildings is one of the main $CO_2$ emission sources in Europe.

As for Italy, most of the residential buildings (77%) were constructed before 1981, when only 25% of the national territory was classified as seismic. These buildings were not designed considering seismic actions; therefore, they are generally characterized by high vulnerability, as clearly shown by recent earthquakes (e.g., L'Aquila 2009, Emilia 2012, central Italy 2016) [2–5].

In addition to seismic deficit, the Italian building stock is also characterized by a large deficit of thermal insulation. In fact, the first regulation, addressing thermal performance criteria, was introduced in 1991 [6], when about 88% of the present Italian building stock had already been realized.

Due to the huge number of buildings having inadequate seismic and energy performances, both in Europe and in Italy, an integrated approach in the design of interventions able to provide multiple beneficial effects is strongly required in view of deploying an effective rehabilitation program. On the contrary, uncoupled rehabilitation solutions appear ineffective. In fact, in earthquake-prone areas, thermal rehabilitation interventions designed neglecting seismic actions could determine an increase of the exposure in terms of building value without adequate seismic protection. Similarly, seismic retrofitting interventions paying little or no attention to thermal rehabilitation could compromise comfort and energy maintenance costs since they can induce higher heat loss (e.g., by simply adding RC shear walls). As an example, interventions that either do not eradicate or—even worse—introduce thermal bridges, due to discontinuities or gaps in the insulation material, can compromise thermal insulation.

In the framework of an integrated approach, the chapter focuses on the impact on the seismic performance of some rehabilitation techniques generally adopted to enhance the thermal performance of infill walls. To this purpose, incremental dynamic analyses (IDAs, [7]) have been carried out on a prototype RC building—representative of the existing building stock—before and after thermal rehabilitation. Specifically, an intervention carried out by replacing the existing masonry infill walls with new elements able to ensure an adequate thermal insulation is considered. Furthermore, the improvement in both seismic and thermal performances through the so-called "double skin" [8, 9] intervention technique, that is, by adding new infilled frames adequately linked to the existing ones, is analyzed.

Energy performances have been investigated following the European standard methods by means of a quasi-steady state approach, based on monthly averaged climate data, and of a dynamic approach, based on hourly climate data. Both energy and seismic analyses have been performed considering two different Italian cities, namely Palermo (Southern Italy) and Milan (Northern Italy), characterized by different climatic conditions and seismic hazard.

## 2. Case study: building type description and modeling

A six-storey RC framed structure representative of the nonseismic post-1971 Italian building stock has been considered as case study. It has a rectangular shape in plan (**Figure 1**) with total dimensions 21.4 × 11.8 $m^2$ (X and Y direction, respectively) and constant interstorey height equal to 3.05 m. As a consequence of the design carried out only for vertical load and the orientation of the one-way RC slab spanning along the longitudinal X direction, frames are arranged in the transversal Y direction only. Rigid beams (0.30 × 0.50 $m^2$) are arranged along the exterior frames,

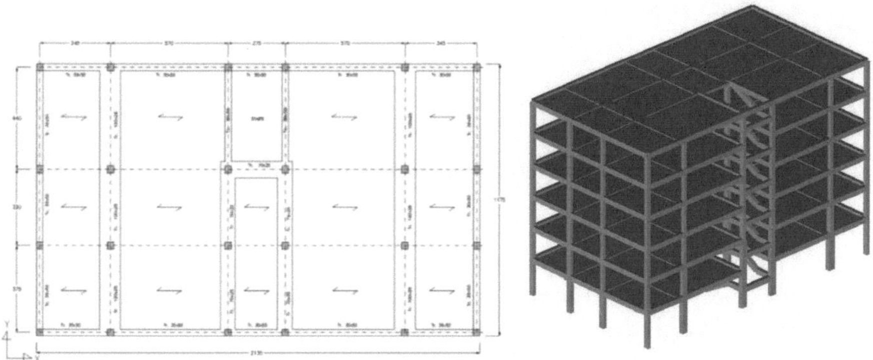

a) b)

**Figure 1.**
*In plan layout of the building type under study (a) and three-dimensional (3D) view of the model (b).*

while internal beams are flexible (0.70 × 0.25 and 1.00 × 0.25 $m^2$). Specifically, along the longitudinal direction X, rigid beams are present only in the exterior frames. Columns have generally cross-sectional dimensions equal to 0.30 × 0.30 $m^2$, except for some columns of the lower storeys whose dimensions range from 0.30 × 0.40 to 0.30 × 0.55 $m^2$. The staircase substructure is placed in a symmetric position with respect to the Y direction and has knee-type beam with dimension 0.30 × 0.50 $m^2$.

According to the considered construction period, infill panel is a multilayer type constituted by two hollow brick walls (8 cm + 12 cm, respectively, internal and external layer) with an air gap (10 cm) and internal/external cement plaster/finish coats (2 cm + 2 cm). Windows are made by wood and a single glass and are characterized by high values of transmittance.

Reinforcement details have been obtained by means of a simulated design procedure [10] with reference to the codes in force and the constructive practice of the period.

Both energy and seismic analyses have been carried out by considering three different configurations:

1. C1: as-built;

2. C2: external existing layer of the as-built infill (12 cm thick) replaced by new one (20 cm thick) with high thermal insulation properties;

3. C3: new 20-cm-thick infill panels placed on new RC frames added to the as-built configuration (C1) and effectively connected to the existing frames.

### 2.1 Energy retrofitting modeling

The energy balance has been performed in the framework of the monthly quasi-steady state and of the hourly dynamic approaches, focusing on the second floor of the building, with $S_u$ = 219.68 $m^2$ of useful floor area, $V$ = 751.96 $m^3$ of volume, and surface-to-volume ratio $S/V$ = 0.6, being $S$ the total external surface.

The boundaries for the calculation of the heating and cooling energy values consist of all the elements separating the conditioned single-zone space from the external environment or unconditioned spaces: external walls and windows. Floors and ceiling are excluded from the envelope representing boundaries with conditioned zones.

The mathematical models adopted to investigate the energy demand of the building are as follows:

1. the simplified monthly quasi-steady state method based on the European Standard ISO 13790 and the UNI 11300:2014, calculating the seasonal energy balance of the building (time interval depending on the climate zone) [11];

2. the hourly dynamic approach based on the recent European Standard ISO 52016 in which a more complex (with respect to the ISO 13790:2008) and an extensive lumped model adopting a resistance-capacitance (RC) network to perform the hourly calculation is considered [12, 13].

The quasi-steady state model includes all sources and sinks of energy, as well as all energy flows through its envelope. The dynamic effects are taken into account by introducing dimensionless utilization factors for heating and cooling [12] that depend on the building inertia by means of the building time constant.

The envelope encloses the volume with a fixed designed temperature for all weather conditions by the use of heating or cooling source energy. Heat flows depend on external and internal influence factors and can be classified as follows:

- transmission losses, $Q_{tr}$, which flow through the building envelope from inside to outside by conduction or heat transfer;

- ventilation losses, $Q_v$, caused by exchange of warm indoor air with colder outdoor air through joints by infiltration or exfiltration, respectively. In addition, room air can be exchanged through open windows or by a mechanical ventilation system. Ventilation is indispensable to assure the hygienically necessary air exchange rate;

- solar gains, $Q_{sol}$, due to the irradiation of solar energy through windows and other transparent or translucent constructional elements. Also added to the solar gains, it is that part of the solar heating of the opaque building envelope, from which the indoor area benefits. Solar heat sources result from the solar radiation normally available in the locality concerned, the orientation of the collecting areas, the permanent shading, the solar transmittance and absorption, and thermal heat transfer characteristics of collecting areas;

- internal gains, $Q_{int}$, represented by the heat released by persons, appliances, computers and other electric devices, as well as from illumination (metabolic heat from occupants and dissipated heat from appliances, dissipated heat from lighting devices, heat dissipated from or absorbed by hot and mains water and sewage systems, heat dissipated from or absorbed by heating, cooling and ventilation systems, heat from or to processes and goods).

The amount of energy, which is necessary to maintain the desired room temperature by compensating the excess of losses ($Q_{tr}$ and $Q_v$) compared to the gains ($Q_{int}$ and $Q_{sol}$), is represented by the energy need for heating, $Q_H$, normalized by the corresponding useful area. To achieve a remarkable reduction in energy need, especially in colder climate zone, renewable energy technologies should be implemented and integrated in the building [14, 15].

Finally, the hourly dynamic model is based on the ISO 52016 [12, 13] in which building elements are modeled by means of lumped parameters. It is worth noting that adopting dynamic models in performing energy analysis, both in design of new buildings and in rehabilitation of existing buildings, can reduce the energy demands for HVAC systems.

### *2.1.1 The effects of thermal bridges*

By following the definition given in the European Standard EN ISO 10211:2017, thermal bridges are part of the building envelope where the otherwise uniform thermal resistance is significantly changed by full or partial penetration of the building envelope by materials with a different thermal conductivity, or a change in thickness of the fabric, or a difference between internal and external areas, such as occurring at wall, floor, and ceiling junctions. They can contribute to increase the energy demand during heating and cooling seasons and can create interior surface condensation problems. Thermal bridges can be classified as follows [11]: (i) repeating thermal bridges, occurring with a regular pattern; (ii) nonrepeating thermal bridges, such as the bridging of a cavity wall by a single lintel; (iii) geometrical thermal bridges, placed at the junction of two planes. Some examples of thermal bridges in building envelopes are corners that provide additional heat flow paths, window sills, floor to wall junctions, balcony supports, lintels, gaps in insulation, and debris in wall cavity.

Thermal bridges produce an additional heat loss affecting the energy balance. It is accounted for by means of the linear thermal transmittance, $\psi$, that is, the rate of heat flow per temperature degree difference per unit length of the junction. In order to reduce heat transmission losses, numerical investigations have been carried out introducing a thermal barrier with different size. In the present chapter, a polystyrene coat of variable size has been added and the effects of thermal insulation have been numerically investigated by using a 2D finite element numerical model.

## 2.2 Seismic modeling

Structural modeling was defined according to a nonlinear macromodeling approach in the OpenSees [16] framework. At both ends of each structural member, a bending moment-rotation (M-δ) relationship was defined by adopting the Ibarra, Medina, and Krawinkler model [17] (**Figure 2**), whose backbone parameters were evaluated according to Haselton and Deierlein [18].

For members having brittle failure, the M-δ relationship was appropriately modified according to the Sezen model [19] in order to account for a lower

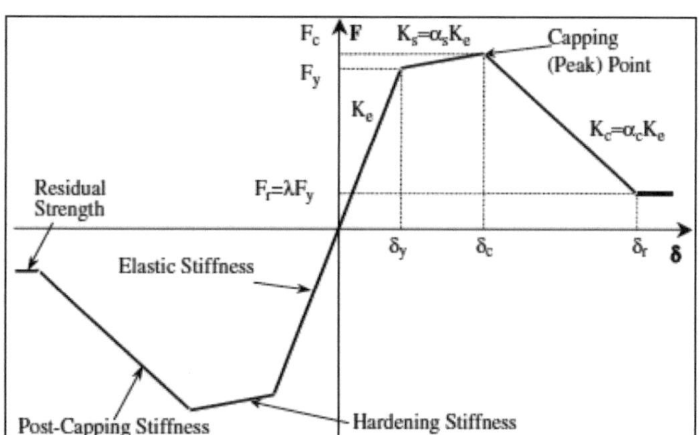


**Figure 2.**
*Adopted backbone curve for structural members (from Ibarra et al. [17]).*

deformation capacity experienced in case of failure in shear. On the basis of the mechanical properties of the constituent materials typically found in real buildings of the period under consideration [20], mean concrete strength value ($f_{cm}$) equal to 20 MPa, and mean steel strength value ($f_{ym}$) equal to 400 MPa were assumed in evaluating the structural capacity.

In the different configurations considered in the study, infill panels were modeled by using a nonlinear equivalent diagonal strut according to the Bertoldi model [21].

Consistent with experimental results (e.g., [22]), for the infill type considered in the as-built configuration, the compressive strength and the elastic modulus are equal to 1.1 and 1800 MPa, respectively. As for infills in both C2 and C3 configuration, the adopted values are 4.0 and 3300 MPa, respectively, for the compressive strength and the elastic modulus, as suggested in [23].

## 3. Thermal and seismic rehabilitation design

The integrated intervention here investigated for the rehabilitation of the RC existing buildings under study consists of the following:

- replacing the external existing infill panel of the as-built configuration (C1) with new one having better thermal insulation properties (C2 configuration);
- adding new infill panels to the as-built configuration (C1) placed on new RC frames effectively connected to the existing ones (C3 configuration).

According to the European Standard EN ISO 6946, 10077, and 12631, the total thermal transmittances for cases C1, C2, and C3 are 0.69, 0.28, and 0.28 W $m^{-2}$ $K^{-1}$, respectively. In dynamic conditions, heat transfer coefficients, including the thermal bridge contribution, have been calculated according to the ISO 13789:2017.

In the following, starting from the assessment results obtained for the as-built configuration (C1), both energy and seismic performances of the two proposed integrated solutions have been evaluated with respect to two different locations, namely Milano and Palermo.

### 3.1 Energy analysis

Numerical simulations have been performed in order to calculate the energy need in configurations C1, C2, and C3. Results have been obtained in both quasi-static and dynamic condition. Two different locations have been considered: (i) Milan, located in the Italian climate zone E (heating degree days in the range 2101–3000, heating period from the 15th of October to the 15th of April and collects about 4000 cities), and (ii) Palermo, located in the Italian climate zone B (heating degree days in the range 601–900, heating period from the 1st of December to the 31st of March and collects about 150 cities).

#### *3.1.1 Thermal bridges and minimization of heat transmission losses*

Preliminary numerical investigations, in the framework of the monthly quasi-steady state model, have been carried out to evaluate the contribution of thermal bridges to the energy balance of the building. Five different cases, labeled (a)–(e) have been considered:

a. as-built (C1 configuration) without thermal bridges;

b. as built (C1 configuration) with thermal bridges;

c. C2 configuration without thermal bridge contribution;

d. C2 configuration with thermal bridge contribution;

e. C2 configuration with thermal bridge contribution and additional thermal barrier with different width, $s$.

The thermal bridges considered in the simulations are pillars, external corners, and roof-to-wall interfaces, as shown in **Figure 3**.

Thermal bridge contributions can be calculated by modeling the corresponding building component connections as input data for a 2D heat flow numerical code. As an example, in **Figure 4**, the temperature fields across the thermal bridge, obtained by means of the 2D heat flow code, are reported. Specifically, simulation refers to the cases (d) and (e), showing the effectiveness of the thermal barrier to minimize heat transmission losses.

Results obtained by numerical simulations are presented in **Table 1**, where heat transmission losses are reported for the cases (a)–(e).

The first conclusion is the importance of accounting for the thermal bridges for an accurate evaluation of heat thermal losses. Comparing results obtained from cases (c) and (d), one can conclude that the contribution of thermal bridges to the calculation of heat transmission losses is close to 40%, thus such contribution cannot be neglected. Therefore, a parametric study to define an effective solution to minimize thermal bridge losses has been carried out. To this end, a thermal barrier made up of a continuous external polystyrene coat has been added, by varying its thickness. **Figure 5** shows that a 4.0 cm coat appears an effective solution, being able to reduce thermal bridge losses up to 35%.

Finally, the influence of thermal bridges is evaluated also by using the dynamic model. As an example, **Figure 6** shows the total heat transmission losses and the time evolution of the temperature of the internal air obtained in Palermo for the configuration C2.

Results show that, in dynamic conditions, the contribution of thermal bridges can influence dramatically the evaluation of transmission losses that can be underestimated up to a factor 2.

### *3.1.2 Energy performances: monthly quasi-steady state and hourly dynamic models*

Numerical simulations have been performed both in quasi-static and dynamic conditions for the two different locations under study, that is, Milano and

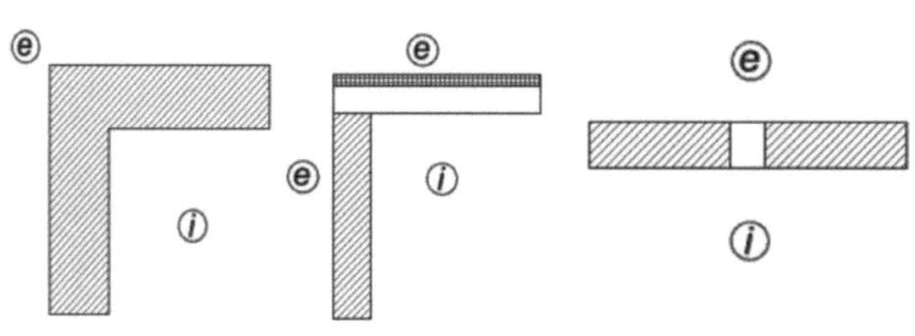


**Figure 3.**
*Thermal bridges: corners (left side), roofs and ceiling (center), and pillars (right side).*

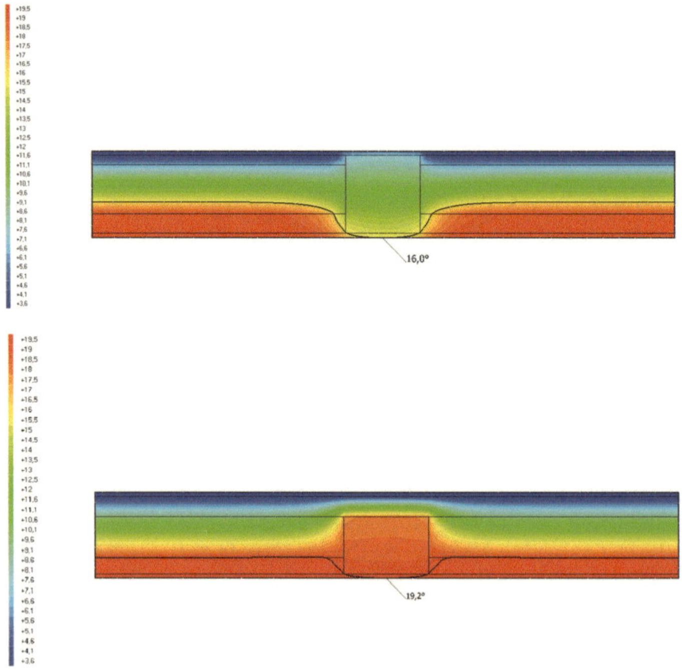


**Figure 4.**
*Thermal bridge simulation of pillars with (on the bottom) and without (on the top) additional 40 mm of thermal barrier.*

| | Case a | Case b | Case c | Case d | Case e (s = 40mm) |
|---|---|---|---|---|---|
| $Q_{tr}$ | 91.76 | 116.21 | 69.78 | 104.05 | 75.44 |
| $Q_{tr}$ pillars | | | | 91.83 | 75.13 |
| $Q_{tr}$ pillars and corners | | | | 93.33 | 75.44 |

**Table 1.**
*Total heat transmission losses (in [*kWh m$^{-2}$ K$^{-1}$*]).*

Palermo. In **Figure 7**, the results computed with the monthly quasi-steady state and hourly dynamic models are compared, both in cooling and in heating conditions. In the simulations, the building volume is divided considering two thermal zones, that is, night and day rooms, with different set point temperatures (19 and 21°C), and the HVAC system is considered active only for 10 hours/day in the case of dynamic simulation. Results show that dynamic and static methods could produce different results due to the capability of the dynamic model of reproducing more realistically the behavior of the building (HVAC system, internal gains or losses,…).

Finally, in **Table 2**, the total thermal energy for heating, $Q_H$, and cooling, $Q_C$, are reported, both for Milano and Palermo by using the hourly dynamic and the monthly quasi-steady state models. Results show the effectiveness of C2 and C3 energy retrofits. In particular, C3 configuration is the best solution both in Palermo and Milano from the point of view of the energy saving. However, it should be highlighted that this is a partial result, as the analysis of the seismic performance, together with the cost of the total retrofit intervention, must be considered in a complete analysis.

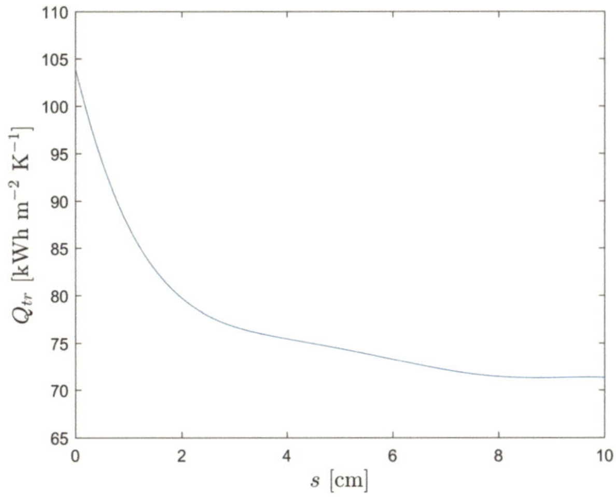


**Figure 5.**
*Total heat transmission losses obtained for the C2 configuration including the effects of thermal bridges and considering an additional thermal barrier.*

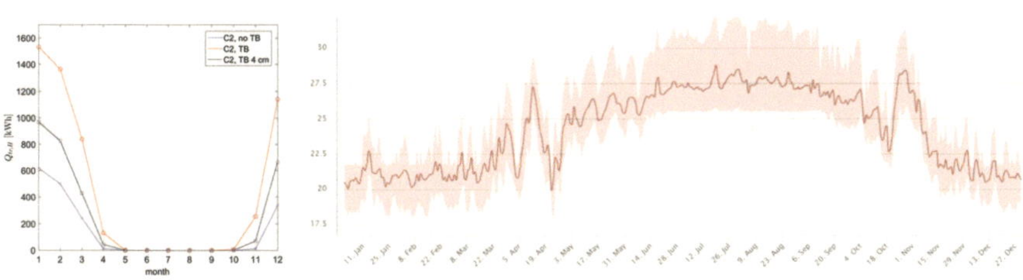


**Figure 6.**
*Total heat transmission losses obtained for the C2 configuration. In the simulations, thermal bridges (TB) and an additional thermal barrier of 4 cm are considered. Simulations are obtained by means of the dynamic model and refer to the building located in Palermo.*

## 3.2 Seismic analyses

Seismic performances have been evaluated through the incremental dynamic analysis (IDA) [7]. In order to account for the effects on the structural response due to record-to-record variability, 10 accelerograms selected in the RINTC project [24] and scaled up to the collapse have been considered. The main seismic parameters of the considered accelerograms are reported in [9] and briefly summarized in **Table 3**.

Results are reported in **Figure 8**. Specifically, the IDA median curves, evaluated for the considered configurations in terms of spectral pseudoacceleration value corresponding to the scaling factor of the record and the maximum base shear value, are displayed separately for X (**Figure 8a**) and Y (**Figure 8b**) in-plane direction. The spectral-pseudoacceleration $Se(T_0)$ has been evaluated at the fundamental period of each considered configuration. In the same figure, the points relevant to both damage limitation (DLLS) and life safety (LSLS) limit state, calculated in accordance with the Italian code [25], are also displayed.

With reference to C1 configuration (i.e., "as-built"), the seismic intensities evaluated in the X direction at DLLS and LSLS are equal to 0.105 and 0.163 g, respectively. In the Y direction, mainly due to the role of the staircase substructure (which determines a greater stiffness with respect to the X direction and a brittle behavior of the relevant short columns), DLLS is achieved at 0.138 g, which is about

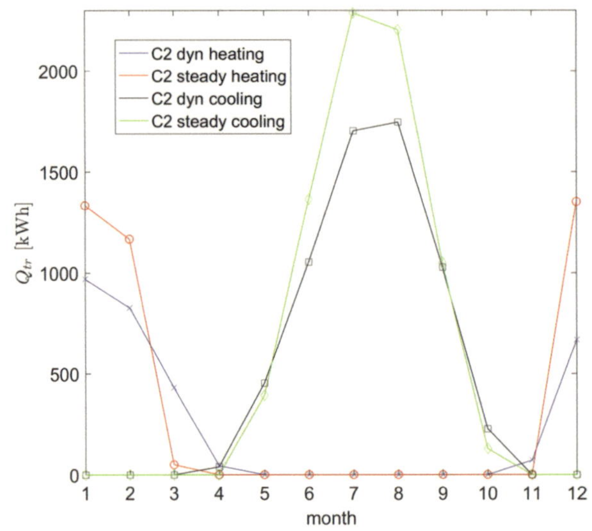


**Figure 7.**
*Total transmission losses (heating and cooling) obtained for the C2 configuration. Simulations are obtained by means of the monthly quasi-steady state and hourly dynamic models and refer to the building located in Palermo.*

| Case | | Milano | | | Palermo | | |
|---|---|---|---|---|---|---|---|
| | | C1 | C2 | C3 | C1 | C2 | C3 |
| Dynamic (kWh) | $Q_H$ | 28,773.69 | 21,078.01 | 9013.55 | 12,627.31 | 8533.53 | 8748.46 |
| | $Q_C$ | 5218.24 | 6345.19 | 6916.87 | 7101.93 | 7795.36 | 8075.09 |
| Quasi-steady state (kWh) | $Q_H$ | 18,665.85 | 14,922.19 | 15,205.30 | 5443.9 | 4129.12 | 4098.47 |
| | $Q_C$ | 3695.57 | 4025.78 | 3809.08 | 7142.80 | 7275.69 | 7401.05 |

**Table 2.**
*Total thermal energy for heating and cooling in Milano and Palermo.*

30% higher than that evaluated in the X direction (0.105 g), whereas a remarkably lower value (0.110 g) than that evaluated in the X direction (0.163 g) is found for LSLS.

In order to estimate the seismic deficit, the above reported intensity values (capacity, $S_{e,C}$) have been compared with the seismic hazard (demand, $S_{e,D}$) evaluated for Milan and Palermo (representative of sites with low and medium seismic hazard) according to the Italian hazard map [26] (soil A). The ratio between the capacity and the demand value ($\alpha = S_{e,C}/S_{e,D}$) for both limit states has been computed, and the results are reported in **Table 4**.

Except for the site of Palermo at LS limit state, both αDL and αLS are always higher than 1 (i.e., no seismic intervention is required). αLS evaluated for Palermo is equal to 0.81, thus asking for a strengthening intervention to guarantee adequate structural safety.

As a consequence of the greater mechanical properties of the infill panels adopted for C2 configuration, a greater base shear value is found with respect to C1 configuration, as shown in **Figure 8**. In terms of seismic intensity values relevant to both DLLS and LSLS, they are equal to 0.130 and 0.197 g in the X direction (to be compared with 0.105 and 0.163 g in the C1 configuration, respectively), while, in

| ID | PGA (g) | PGV (cm/s) | $S_{e,max}$ (g) | HI (m) |
|---|---|---|---|---|
| 1 | 0.046 | 1.61 | 0.167 | 0.049 |
| 2 | 0.035 | 2.13 | 0.125 | 0.054 |
| 3 | 0.043 | 0.41 | 0.183 | 0.054 |
| 4 | 0.021 | 0.83 | 0.086 | 0.052 |
| 5 | 0.070 | 0.26 | 0.339 | 0.047 |
| 6 | 0.050 | 2.37 | 0.089 | 0.052 |
| 7 | 0.029 | 0.45 | 0.108 | 0.061 |
| 8 | 0.009 | 0.85 | 0.037 | 0.083 |
| 9 | 0.026 | 0.31 | 0.105 | 0.064 |
| 10 | 0.028 | 0.56 | 0.147 | 0.055 |

**Table 3.**
*Seismic parameters of the considered accelerograms in terms of peak ground acceleration (PGA), peak ground velocity (PGV), maximum spectral pseudoacceleration ($S_{e,max}$), and Housner intensity (HI).*

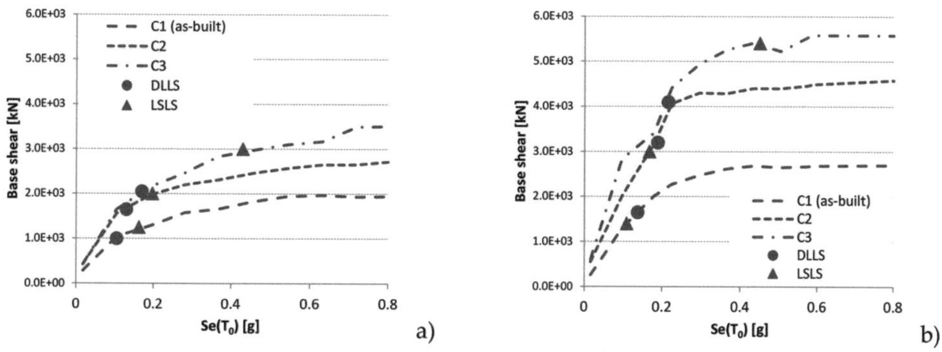


**Figure 8.**
*Spectral pseudo-acceleration versus maximum base shear curves relevant to the three considered configurations, obtained for X (a) and Y (b) direction by considering median values from IDA analyses.*

| | DLLS | | | LSLS | | |
|---|---|---|---|---|---|---|
| **Site** | **Se,C (g)** | **Se,D (g)** | **αDL = Se,C/Se,D** | **Se,C (g)** | **Se,D (g)** | **αLS = Se,C/Se,D** |
| Milan | 0.105 | 0.015 | 7.0 | 0.110 | 0.042 | 2.6 |
| Palermo | | 0.039 | 2.7 | | 0.135 | 0.81 |

**Table 4.**
*α ratio values between seismic capacity Se,C and demand Se,D evaluated for both DL and LS limit state.*

the Y direction, they are 0.189 and 0.168 g (0.138 and 0.110 g in the C1 configuration, respectively). Consequently, the minimum αLS value evaluated for the site of Palermo goes from 0.81 to 0.93 (**Table 5**).

As for C3 configuration (i.e., new infilled RC frames added to the as-built configuration C1 and effectively connected to the existing frames, **Figure 9**), **Figure 8** summarizes the results obtained from the IDA analyses and the comparison with the other configurations (C1 and C2). In terms of seismic capacity evaluated at the two considered limit states, the values at LSLS are equal to 0.430 and 0.450 g for X and

| | C1 ("as-built") | | | C2 | | | C3 | | |
|---|---|---|---|---|---|---|---|---|---|
| **Site** | **Se,C (g)** | **Se,D (g)** | **α = Se,C/ Se,D** | **Se,C (g)** | **Se,D (g)** | **α = Se,C/ Se,C** | **Se,C (g)** | **Se,D (g)** | **α = Se,C/ Se,C** |
| Milan | 0.110 | 0.042 | 2.60 | 0.168 | 0.057 | 2.95 | 0.430 | 0.063 | 6.82 |
| Palermo | | 0.135 | 0.81 | | 0.180 | 0.93 | | 0.200 | 2.15 |

**Table 5.**
*Comparison between seismic capacity values and hazard demand relevant to as-built and postintervention configurations evaluated for LS limit state.*

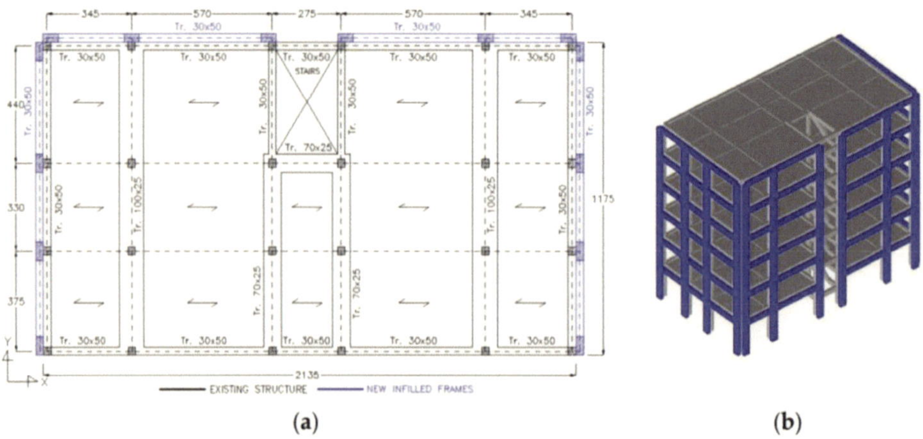


**Figure 9.**
*C3 configuration: in plan layout of the retrofitted building (a) and 3D view of the model (b).*

Y directions, respectively, while for DLLS, intensity values are 0.170 and 0.215 g for X and Y directions, respectively. Consequently, a full seismic rehabilitation (i.e., αLS ratio greater than 1) has been achieved also for Palermo, as reported in **Table 5**.

## 4. Conclusions

In the present work, an application of integrated rehabilitation intervention on reinforced concrete (RC) existing buildings has been presented. A case study located in two different cities, Milano and Palermo, having different climatic conditions and seismic hazard values, has been investigated.

The building under study has been analyzed considering three different configurations, with different arrangement and features of the infill panels, in order to highlight their role on both thermal and seismic performances. Quasi-static and dynamic methodologies have been used for the calculation of the energy demand, highlighting the importance of accounting for thermal bridges in the investigations. As for seismic performance, the results of the incremental nonlinear dynamic analyses show that infill panels having greater thermal and mechanical properties increase the seismic capacity. Nevertheless, a full seismic rehabilitation for the site with the highest seismic hazard (i.e., Palermo) has been achieved only by strengthening the RC structure with additional RC frames.

Future work, besides including additional case studies (building types, different seismic and climate sites,...), will be devoted to set up and propose a methodology for integrated rehabilitation interventions. Further, economic aspects will be better investigated with the aim of finding the cost-optimal integrated retrofit solution including the impact of the expected economic losses due to seismic damage throughout the building life cycle.

## Author details

Antonio D'Angola*, Vincenzo Manfredi, Angelo Masi and Marianna Mecca
Scuola di Ingegneria, Università della Basilicata, Potenza, Italy

*Address all correspondence to: antonio.dangola@unibas.it

## References


[1] Buildings Performance Institute Europe (BPIE). Europe's Buildings under the Microscope. A Country-By-Country Review of the Energy Performance of Buildings. Brussels, Belgium: Buildings Performance Institute Europe; 2011

[2] Braga F, Manfredi V, Masi A, Salvatori A, Vona M. Performance of nonstructural elements in RC buildings during the L'Aquila, 2009 earthquake. Bulletin of Earthquake Engineering. 2011;**9**:307-324. DOI: 10.1007/s10518-010-9205-7

[3] Ricci P, De LF, Verderame GM. 6th April 2009 L'Aquila earthquake, Italy: Reinforced concrete building performance. Bulletin of Earthquake Engineering. 2011;**9**:285-305. DOI: 10.1007/s10518-010-9204-8

[4] Manfredi G, Prota A, Verderame GM, De Luca F, Ricci P. 2012 Emilia earthquake, Italy: Reinforced concrete buildings response. Bulletin of Earthquake Engineering. 2014;**12**:2275-2298. DOI: 10.1007/s10518-013-9512-x

[5] Masi A, Chiauzzi L, Santarsiero G, Liuzzi M, Tramutoli V. Seismic damage recognition based on field survey and remote sensing: General remarks and examples from the 2016 Central Italy earthquake. Natural Hazards. 2017;**86**(1):193-195. DOI: 10.1007/s11069-017-2776-8

[6] Norme per l'attuazione del Piano energetico nazionale in materia di uso razionale dell'energia, di risparmio energetico e di sviluppo delle fonti rinnovabili di energia. Gazzetta Ufficiale n. 13, Law n. 10, 9 January 1991. (16-1-1991—Suppl. Ordinario n. 6. in Italian)

[7] Vamvatsikos D, Cornell CA. Incremental dynamic analysis. Earthquake Engineering and Structural Dynamics. 2002;**31**(3):491-514

[8] Marini A, Feroldi F, Belleri A, Passoni C, Preti M, Giuriani E, et al. Coupling energy refurbishment with structural strengthening in retrofit interventions, Alessio Caverzan. In: Tornaghi ML, Negro P, editors. Proceedings of SAFESUST Workshop. 2015. DOI: 10.2788/499080

[9] Manfredi V, Masi A. Seismic strengthening and energy efficiency: Towards an integrated approach for the rehabilitation of existing RC buildings. Buildings. 2018;**8**:36. DOI: 10.3390/buildings8030036

[10] Masi A. Seismic vulnerability assessment of gravity load designed R/C frames. Bulletin of Earthquake Engineering. 2003;**1**(3):371-395

[11] UNI 11300-1:2014—Energy performance of buildings—Part 1: Evaluation of energy need for space heating and cooling. Milan, Italy: UNI; 2014

[12] ISO 52016-1:2017, Energy performance of buildings—Energy needs for heating and cooling, internal temperatures and sensible and latent heat loads— Calculation procedures. 2017

[13] ISO/TR 52016-2:2017, Energy performance of buildings—Energy needs for heating and cooling, internal temperatures and sensible and latent heat loads—Part 2: Explanation and justification of ISO 52016-1 and ISO 52017-1. 2017

[14] Spertino F, D'Angola A, Enescu D, Di Leo P, Fracastoro GV, Zaffina R, Thermal-electrical model for energy estimation of a water cooled photovoltaic module, Solar Energy. 2016;**133**(1):119-140. DOI: 10.1016/j.solener.2016.03.055

[15] D'Angola A, Zaffina R, Enescu D, Di Leo P, Fracastoro G, Spertino F.

Proceedings of the 51st International Universities' Power Engineering Conference. Coimbra, Portugal: Institute of Electrical and Electronics Engineers Inc., 2016. ISBN: 978-1-5090-4650-8. DOI: 10.1109/UPEC.2016.8114086

[16] McKenna F. OpenSees: A framework for earthquake engineering simulation. Computing in Science & Engineering. 2011;**13**(4):58-66

[17] Ibarra LF, Medina RA, Krawinkler H. Hysteretic models that incorporate strength and stiffness deterioration. Earthquake Engineering and Structural Dynamics. 2005;**34**(12):1489-1511

[18] Haselton CB, Deierlein GG. Assessing seismic collapse safety of modern reinforced concrete moment frame buildings. Blume Report No. 156, 2007

[19] Sezen H. Seismic response and modeling of reinforced concrete building columns [PhD dissertation]. Berkeley: Department of Civil and Environmental Engineering, University of California; 2002

[20] Masi A, Digrisolo A, Santarsiero G. Concrete strength variability in Italian RC buildings: Analysis of a large DataBase of Core tests. Applied Mechanics and Materials. 2014;**597**:283-290. DOI: 10.4028/www.scientific.net/AMM.597.283

[21] Bertoldi SH, Decanini LD, Gavarini C. Telai tamponati soggetti ad azioni sismiche, un modello semplificato, confronto sperimentale e numerico. In: Proceedings of the VI ANIDIS Conference "L'ingegneria Sismica in Italia"; 13-15 October 1993; Perugia. 1993 (in Italian)

[22] Calvi GM, Bolognini D. Seismic response of reinforced concrete frames infilled with weakly reinforced masonry panels. Journal of Earthquake Engineering. 2001;**5**(2):153-185

[23] Guidi G, da Porto F, Dalla Benetta M, Verlato N, Modena C. Comportamento sperimentale nel piano e fuori dal piano di tamponamenti in muratura armata e rinforzata. In: Proceedings of the ANIDIS Conference "L'ingegneria Sismica in Italia"; 30 June–4 July 2013; Padova. 2013 (in Italian)

[24] Iervolino I, Spillatura A, Bazzurro P. RINTC project: Assessing the (implicit) seismic risk of code-conforming structures in Italy. In: Papadrakakis M, Fragiadakis M, editors. 6th ECCOMAS Thematic Conference on Computational Methods in Structural Dynamics and Earthquake Engineering (COMPDYN 2017); 15-17 June 2017; Rhodes Island, Greece. Paper 17282. 2017

[25] Ministrial Decree 17 January 2018, NTC 2018. Aggiornamento delle Norme Tecniche per le Costruzioni, Ministero delle Infrastrutture. Supplemento ordinario alla Gazzetta Ufficiale n. 42 del 20 febbraio 2018—Serie generale (in Italian)

[26] Presidential Decree n. 3519/2006. Criteri generali per l'individuazione delle zone sismiche e per la formazione e l'aggiornamento degli elenchi delle medesime zone. Gazzetta Ufficiale n. 108, 11 maggio 2006 (in Italian)

Chapter 4

# A Smart Battery Management System for Photovoltaic Plants in Households Based on Raw Production Forecast

*Filippo Spertino, Alessandro Ciocia, Paolo Di Leo, Gabriele Malgaroli and Angela Russo*

## Abstract

A basic battery management system (BMS) permits the safe charge/discharge of the batteries and the supply of loads. Batteries are protected to avoid fast degradation: the minimum and maximum state-of-charge (*SOC*) limits are not exceeded and fast charge/discharge cycles are not permitted. A more sophisticated BMS connected to a photovoltaic (PV) generator could also work with the double purpose of protecting storage and reducing peak demand. Peak reduction by storage generally requires the forecast of consumption and PV generation profiles to perform a provisional energy balance. To do it, it is required to have accurate information about production profiles, that is, to have at disposal accurate weather forecasts, which are not easily available. In the present work, an efficient BMS in grid-connected PV plants for residential users is described. Starting from raw 1-day ahead weather forecast and prediction of consumption, the proposed BMS preserves battery charge when it is expected high load and low PV production and performs peak shaving with a negligible reduction in self-sufficiency.

**Keywords:** battery management systems, photovoltaic system, storage, self-sufficiency, peak shaving

## 1. Introduction

Recently, the installation of renewable energy systems (RES), such as photovoltaic (PV) generators, has increased due to dedicated policies and even lower investment costs [1]. The increasing share of RES has introduced new challenges, which in future can affect the proper operation of the system: for example, the decrease of the system inertia faced by introducing new inverter controls [2]. At distribution system level, the large share of RES led researcher to consider new way to manage the system, by means of optimal reconfiguration procedure based on different methodology [3] and time periods [4]. The main drawback of RES is the intermittency of power production, which often results in a not well match between electric consumption and generation profiles [5, 6], with consequent voltage deviations and reverse power flow issues [7]. In order to reduce voltage deviations, it is possible to upgrade grid lines and transformers, but it is generally expensive. Otherwise, it is possible to reduce injection from renewable sources by increasing the supply of local loads. It can be done by load shifting, which consists of the switching on of home appliances, when PV generators are working. This procedure can be manual by using simple timed switches; for example, the user has to switch on the washing machine or the dishwasher at midday, when the production is maximum. To perform load shifting, in [8], it is developed an algorithm to predict the consumption based on hourly historical data using artificial neural networks (ANNs).

Using electrochemical storage with PV generators is a good alternative to mitigate or eliminate power injection issues. Storage is easy to install and manage in any site; in the last years, the cost of storage decreases, but it is still expensive and it cannot solve the seasonal correlation between low loads and high RES production, and vice versa. For this reason, in case of domestic users, the best technical-economic solution is the use of a small battery system (BS) and the adoption of load shifting. This solution permits the reduction of absorption or injection peaks and the increase of self-sufficiency level, that is, the ratio between the local RES production used to supply loads and the total loads.

A battery management system (BMS) is a hardware/software solution which checks the correct operation of batteries: in its basic version, it simply charges the batteries, when they are empty, and discharges them when necessary. It limits battery operation only to protect them: the exceeds of minimum and maximum state-of-charge (*SOC*) limits and fast charge/discharge cycles are not permitted to avoid fast degradation [9]. An improvement in the BMS management consists of the forecast of load and PV generation profiles [10, 11]. In this case, it is necessary to have accurate information about production profiles, which are generally missing. In addition, the BMS has to continuously obtain accurate weather forecasts, which are not easily available. In [12], a modified control strategy for batteries based on peak shaving is proposed to reduce power fluctuations of production in a PV-storage system and obtain benefits in terms of electricity price. In [13], a more accurate BMS for a PV-storage system is developed: the proposed management strategy aims to shave consumptions peaks, taking into account degradation of batteries and aging limits of the storage. A real-time battery management algorithm is proposed in [14] to reduce the peak demand power and the daily energy cost in grid-connected PV-storage systems. In particular, the charge/discharge of the storage is controlled using instantaneous load data. Each day, 1-day ahead prediction of PV generation and load profiles is performed to decide the power limit beyond which the peak shaving strategy works. Finally, in [15], several control strategies of batteries are compared for a residential battery energy storage system (BESS) coupled with a PV generator. In particular, a base control strategy charges the battery when PV production exceeds local loads and starts to discharge the storage in the evening, when PV generation is negligible. It is compared to three optimized BMSs: the first one aims to maximize the economic benefits for the users or the self-sufficiency, while the second one includes utility constraints to lower overvoltage risks on distribution grid and the third is a distributed control.

In the present chapter, positive aspects regarding the grid stability, i.e., frequency and voltage control [16–18], are not taken into account, and only the benefit for the users, consisting of the reduction of absorption peaks with a possible consequent reduction of contracted power, is investigated. In addition, load shifting is not considered, due to difficulties in convincing domestic users to change their habits. Indeed, a smart battery management system (SBMS), which works with raw forecasts of production and historical consumption data, is proposed: the goal of the control is to reduce the absorption peaks from the grid with minimum reduction in self-sufficiency and no load shifting. In particular, in case of high consumption and low production, a traditional BMS completely discharges the batteries and all the renewable energy is locally consumed. In the proposed SBMS, the storage will not be totally discharged and will not completely supply the loads. In fact, the storage discharge is limited to satisfy possible absorption peaks in a period up to few days. Nevertheless, if the storage is not discharged waiting for possible consumption peaks, it means that the baseload could not be satisfied with a consequent reduction of self-sufficiency. The self-sufficiency is calculated to check the effectiveness of the proposed SBMS: the domestic user has to keep high its self-sufficiency level, because it corresponds to an economic return. The benefit for the grid is not taken into account, but it exists: it consists of a reduction in peak absorption from the grid resulting in higher power quality, lower voltage dips, and reverse power flow issues [19, 20].

The next sections of the chapter will be organized in the following way. In Section 2, the description of the system setup, the inputs for the simulation, and the models of the PV generator and the battery will be presented. In Section 3, the provisional energy balance and the storage management are described in detail. In Sections 4 and 5, the results of the simulations and the conclusions are discussed, respectively.

## 2. The simulated PV-storage system

### 2.1 Description of the system

A scheme of a PV-storage residential system is presented in **Figure 1**. The main components of the power system are a PV generator, an electrochemical BMS, DC/DC and DC/AC power converters, AC loads, and the distribution grid. The PV modules are connected to a maximum power point tracker (MPPT) in order to work in the maximum power point in every irradiance and temperature condition [21]. The BMS measures DC current and voltage and temperature of batteries. The *SOC* is continuously calculated in order to estimate the residual charge of the storage; in this way, the BMS avoids an abnormal degradation of the batteries due to not optimal charging patterns, overcharging, undercharging, and abnormal temperatures.

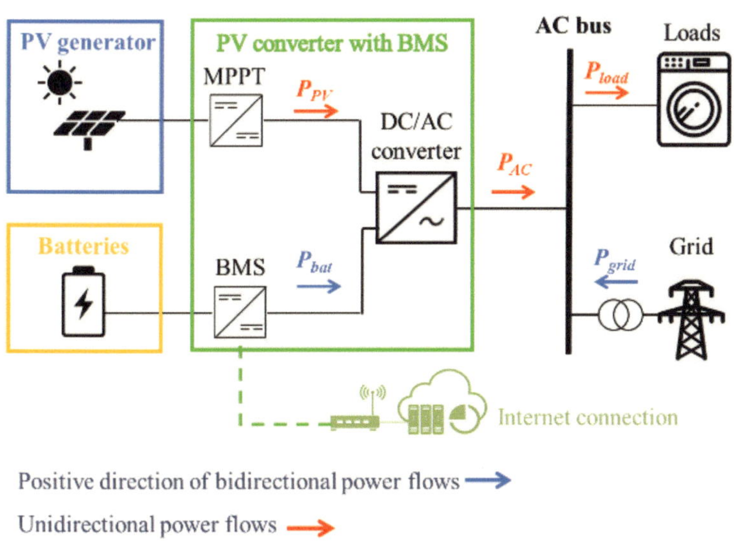


**Figure 1.**
*The PV-storage system under study.*

The DC/AC converter connects the PV system and the BESS to the AC side, i.e., local loads and the grid. Moreover, the device is Internet-connected and downloads raw weather forecast of 1-day ahead, compares provisional load and production profile, and adopts the best strategy to reduce consumption peaks.

### 2.2 Usage of PVGIS for estimation of irradiance profiles

The production of PV generators depends on installation conditions (location and tilt and azimuth of the PV modules) and on weather conditions (solar irradiance and temperature). The Photovoltaic Geographical Information System (PVGIS) [22] database is a free online tool; it permits to know the average daily irradiance and temperature profiles corresponding to each month of the year. The monthly profiles can be obtained for every location in Europe, Africa, and Asia starting from the definition of the location of the generator and the tilt and the azimuth of the PV modules. Additional parameters can be selected, such as the typology of solar radiation database and the calculation of irradiation profiles, also for tracking systems. In **Figure 2**, the home screen of PVGIS database is shown.

The PVGIS database provides a temperature profile and three irradiance profiles for each month. In particular, the irradiance profiles correspond to a clear sky day, an average day, and an overcast day. During the clear sky day, the global irradiation is maximum; in fact, it is mainly composed of the beam contribution, because no clouds are present. During the overcast day, the solar irradiation is minimum; in fact, in case of cloudy and rainy days, only the diffuse component of the solar irradiance is present. The average day is an intermediate situation: it is based on the average irradiance condition occurring in the month under consideration. In **Figure 3**, an example of the output profiles of the software is presented for January; the selected location is in Italy (Turin, 45.05° Nord, 7° 40' Est) and the PV modules are installed with an inclination of 15° and West oriented (azimuth = 90°, where South = 0°). Data are provided with a time step of 15 min.

In the present chapter, it is supposed to install a single device including both the PV converter and the BMS; the BMS will be equipped with additional hardware and software capable of accessing Internet and download data from the PVGIS database. After the installation of the PV generator, during the setting up of the converter, the input parameters requested by PVGIS to estimate the irradiance and temperature profiles are inserted in the software of the device.

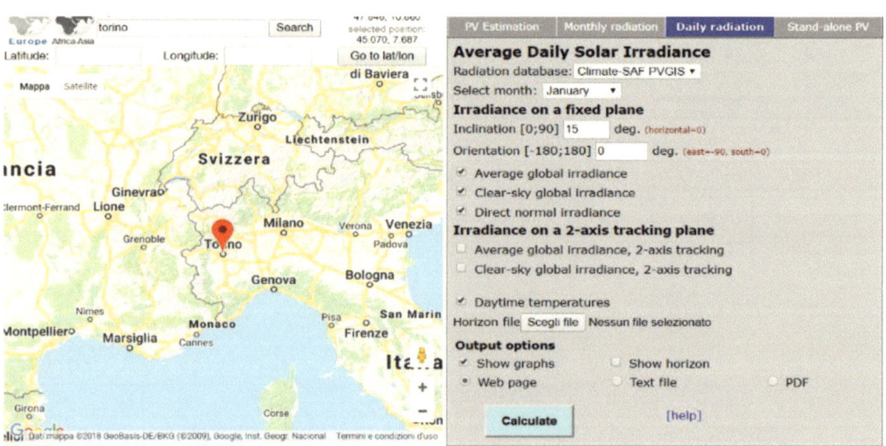


**Figure 2.**
*PVGIS website.*

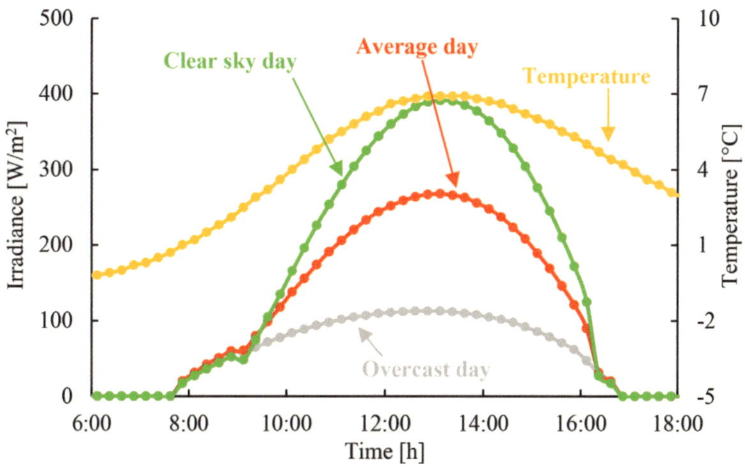


**Figure 3.**
*Irradiance and temperature profiles for January in Turin (Italy) from PVGIS database; PV modules have inclination of 15° and West orientation.*

The device accesses the PVGIS database and downloads and elaborates the three above-described irradiation profiles for each month. Starting from these data and the rated power of the PV generator, the power converter calculates a total of 36 PV production profiles by an appropriate photovoltaic model, which will be described in detail in Section 2.3. Finally, the power generation profiles are integrated over the entire day: the result is a list of daily energy productions for each month in three different weather conditions. **Table 1** shows the daily energy production of a PV generator with rated power of 1 $kW_p$ installed as defined in **Figure 3**.

## 2.3 Modeling of PV generators

Regarding the PV power simulation, the AC power production $P_{AC}$ is calculated according to the model described in [23]. The inputs of the model are solar irradiance $G$, ambient temperature $T_a$, and rated power of the PV generator $P_{PV,r}$.

| PV production (kWh/kW$_p$) | Clear sky day | Average day | Overcast day |
|---|---|---|---|
| January | 2.0 | 1.5 | 0.8 |
| February | 3.0 | 2.5 | 1.0 |
| March | 4.3 | 3.7 | 1.6 |
| April | 5.6 | 4.3 | 1.8 |
| May | 6.4 | 5.2 | 2.2 |
| June | 6.6 | 5.6 | 2.2 |
| July | 6.5 | 5.8 | 1.9 |
| August | 5.8 | 5.0 | 1.8 |
| September | 4.6 | 3.9 | 1.5 |
| October | 3.3 | 2.5 | 1.3 |
| November | 2.2 | 1.6 | 0.8 |
| December | 1.8 | 1.3 | 0.7 |

**Table 1.**
*Daily energy production for each month in three different weather conditions.*

The thermal losses and consequently the DC input power change, while the other sources of losses are considered constant:

$$P_{AC} = P_{PV,r} \cdot \frac{G}{G_{STC}} \cdot \eta_{mix} \cdot \eta_{therm} \cdot \eta_{DC/AC} \tag{1}$$

with

$$\eta_{mix} = \eta_{dirt} \cdot \eta_{refl} \cdot \eta_{mis} \cdot \eta_{MPPT} \cdot \eta_{cabl} \cdot \eta_{shad} \tag{2}$$

Losses due to temperature $\eta_{therm}$ are due to the reduction in the voltage of the PV generation with increasing temperature. The power loss with respect to the standard test condition is linearly dependent on temperature with proportionality factor about $\gamma_{th} \approx 0.3 \div 0.5\%/°C$ depending on the semiconductor of the photovoltaic generator [24]. According to [25], a value of 0.5%/°C, typical for c-Si PV modules, which is the most diffused PV technology for terrestrial applications, is used. In order to estimate temperature losses, at every time step, the temperature of the PV cells $T_c$ is calculated starting from measured air temperature $T_a$ by the following equation [26]:

$$T_c = T_a \cdot \frac{NOCT - 20^{\circ}C}{G_{NOCT}} \cdot G \tag{3}$$

*NOCT* is the normal operating cell temperature, generally provided by the manufacturer of the PV modules; in this work, it corresponds to a typical value *NOCT* = 45°C. $G_{NOCT}$ is the solar irradiance occurring at *NOCT* condition and it is 800 W/m$^2$; the overtemperature losses $\eta_{therm}$ (with respect to $T_{STC}$ = 25°C) are calculated by the formula:

$$\eta_{therm} = 1 - \gamma_{th} \cdot (T_c - T_{STC}) \tag{4}$$

Losses due to dirt $\eta_{dirt}$ provide an average 2% of reduction in energy production for the deposit of dust and other materials on the glass of the modules. Thus, a typical value $\eta_{dirt} = 0.98$ is used in the present work [27, 28]. Note that in case of horizontal modules, the cleaning made by rain is reduced, and in case of emission of pollution close to the plants from special industrial processes [29], losses can be more than 7%. Losses due to reflection from the glass on the front of the PV module are inevitable losses due to a not ideal transparency of the glass; according to [30], they can be considered equal to ≈3% ($\eta_{refl} = 0.97$). Mismatch losses are due to nonuniformity in *I-V* characteristics of modules connected in series or in parallel. Thus, the conversion unit imposes to the whole PV generator a working point not perfectly corresponding to the optimum. According to [31], a typical value of $\eta_{mis} = 0.97$ is used. Joule losses take into account dissipation of electrical energy into heat by Joule effect in the cables. During design phase, cables should be sized in order to keep Joule losses within 3% in nominal conditions [30]. Since the PV system operates at nominal conditions (maximum power) for a short period during the year, and in other conditions (partial load) losses are lower, Joule losses are estimated equal to an average value of 1% ($\eta_{cabl} = 0.99$) [30]. Losses for shadings are due to external causes, as a wrong design; thus, in the performed simulations, these losses are neglected ($\eta_{shad} = 1$). Regarding the accuracy of the maximum power point tracking (MPPT) system, it causes losses, because the optimum value is generally not perfectly tracked, especially at low power: on average, this loss can be estimated ≈1%. ($\eta_{MPPT} = 0.99$) [32]. Finally, the DC/AC conversion introduces losses, which are quadratically dependent on the power output. For the sake of simplicity, an average value of $\eta_{DC/AC}$ = 0.97 is considered in the present work [33].

### 2.4 Modeling of electrochemical storage

A correct model of the storage is fundamental to evaluate energy flows. Many electric models are present in literature and they permit to simulate operation of batteries with different pros and cons [34–37]. The simplest model describes a battery by an equivalent voltage source in series with an internal resistance. The equivalent voltage can easily be determined by measuring the open circuit voltage of the battery, while the measurement of the internal resistance requires a further test performed during battery charge. Obviously, this model has a limited use, because the parameters are constant: the accumulator results in having an infinite capacity and there is no way to determine the *SOC*. An upgrade with respect to the basic model is obtained using an equivalent resistive-capacitive model [36]. The values of resistances and capacitances can be determined through impulsive test of the battery. The advantage of this model is that it permits to evaluate the charge and discharge transients with variable loads in time. However, the SOC dependence on the voltage, which has to be determined, requires careful preliminary measurements on the battery. Another possible model consists of the impedance model, where a voltage source is in series with a resistance and an inductance. An additional series impedance is used to represent the electrochemical characteristics of the battery. Nevertheless, the definition of this impedance is complicated; in fact, it can be obtained starting from an electrochemical impedance spectroscopy to obtain an equivalent impedance in the frequency domain. In addition, the impedance has to be characterized varying the state of charge and the temperature [35].

The most sophisticated models [38, 39] are developed to calculate also the state of health (*SOH*), which is a parameter useful to evaluate how the charge-discharge profiles affect the storage life and when the batteries have to be replaced. In fact, PV production is intermittent; thus, PV generators cannot guarantee the optimal charge-discharge cycles to have the longest possible life of storage and the highest efficiency. For example, the *real-time model* described in [39] is a blend of the previous battery models whose particular combination of components and dependencies eases the estimation of the equivalent parameters. In conclusion, this model permits to calculate the *SOC*, the *SOH*, and then the residual life. This information permits the evaluation of the economic investment of electrochemical storage system [40], taking in consideration the battery management. Nevertheless, the models that permit to estimate the *SOH* require a continuous measurement of battery parameters (i.e., voltage, current, and temperature of the batteries) [37]. For this reason, the calculation of the *SOH* cannot be performed with only simulations, but a real system with continuous measurements is required.

The *energy model* is used in the simulations presented in this chapter, because it permits to simulate the *SOC* with a good approximation (a few percent points) without measurements and with a low computation effort (only the formulas (5) and (6) are used). The *energy model* permits to estimate the state of charge of batteries; i.e., how much energy is stored or can be stored in a battery with rated energy capacity $C_{bat}$, by the comparison with the limits imposed to preserve life of batteries. The calculation of the $SOC(t)$ at the instant $t$ is a function of the state of charge $SOC(t-1)$ at the previous time step, of the power exchanged $P_{\text{bat}}$ during the time step $\Delta t$ (in this chapter, $\Delta t$ = 1 min) and of the charge efficiency $\eta_{bat}$. During the charge phase, the batteries behave as a generator ($P_{\text{bat}} > 0$) and it is considered a charge efficiency $\eta_{bat}$ = 0.88; during discharge ($P_{\text{bat}} < 0$), efficiency is considered unitary.

$$SOC(t) = SOC(t-1) + \frac{\eta_{bat} \cdot P_{bat} \cdot \Delta t}{C_{bat}} \quad P_{bat} > 0 \tag{5}$$

$$SOC(t) = SOC(t-1) + \frac{P_{bat} \cdot \Delta t}{C_{bat}} \quad P_{bat} < 0 \tag{6}$$

## 3. Provisional energy balance and storage management

The proposed BMS periodically defines the strategy to minimize the power absorption from the grid. The strategy selection is performed two times per day to better match the consumption peaks of domestic users, which occur early in the morning and during the evening. Thus, the day is divided in three time slots. The first time slot starts at midnight and ends at 6:00 a.m. Between 6:00 a.m. and 6:00 p.m., there is the second time slot: the production is dominant and in case of people at home, part of generation is self-consumed. In this period, the consumption peak in the morning due to preparation to work and school activity (such as hairdryers, electric boiler, etc.) is included. Obviously, this peak cannot be totally satisfied by PV production, especially in winter. The third time slot starts at 6:00 p.m. and finishes at midnight, when the second consumption peak occurs, and PV production is low or negligible.

The time 6:00 p.m. is selected for the download of raw weather forecasts for the next 24 h, for the calculation of provisional energy balance and the update of management strategy for batteries. In fact, at 6:00 p.m., the PV production is almost over: the BESS can accurately calculate the quantity of stored energy, which will be available for the next hours. In fact, during evening and night, the batteries will not be charged: supply from the grid is not considered.

### 3.1 Comparison of estimated energy production and consumption

The provisional energy balance for 1-day ahead is performed comparing estimated energy production and consumption. Regarding the energy consumption, this value is calculated on the basis of measurement of local consumption profiles. Loads are monitored, and average values of energy consumption are calculated for each of the three time slots composing the day, as described in the previous paragraph. In addition, a distinction of average energy consumption between working days and holidays is considered.

Regarding the provisional production, every day at 6:00 p.m., the converter downloads raw weather forecasts for the next 24 h. Data are collected from commercial web services: they generally identify weather forecast with simplified symbols, i.e., showing a sun symbol for a clear sky day and lightning for rain. For the sake of simplicity, in the present work, it is considered a three-level forecast: a clear sky day, an average day with few clouds, and a cloudy/rainy day. These levels correspond to the three irradiance conditions provided by the database PVGIS. In this way, it is defined a raw correlation between the weather forecast and the expected production from the PV generator. The advantage consists of a free and easily accessible daily forecast of production, which can be used for free by the Internet-connected BESS to select the best battery management.

### 3.2 Definition of the total discharge time

The first step in the smart management of batteries consists of the definition of the total discharge time (*TDT*): **Figure 4** shows the flowchart of the procedure.

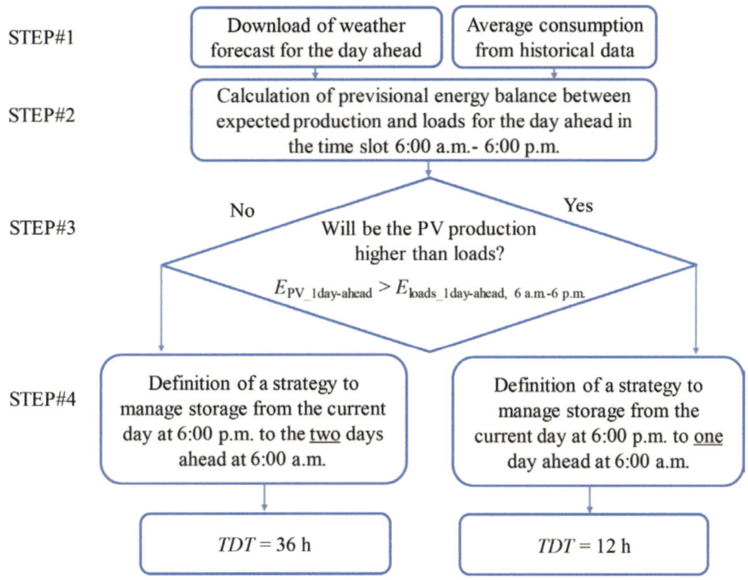


**Figure 4.**
*Definition of the total discharge time (TDT).*

First, at 6:00 p.m., after weather forecast download, the provisional balance between expected production $E_{PV\_1day\text{-}ahead}$ and loads $E_{loads\_1day\text{-}ahead,\ 6\ a.m.\text{-}6\ p.m.}$ occurring in the time slot 6:00 a.m.–6:00 p.m. of the day ahead is performed.

In case of PV energy production higher than loads $E_{PV\_1day\text{-}ahead} > E_{loads\_1day\text{-}ahead,\ 6\ a.m.\text{–}6\ p.m.}$, a management of the storage to satisfy loads until 1-day ahead at 6:00 a.m. is performed. In this case, the *TDT* will be equal to 12 h. In fact, the day after, during light hours, energy will be self-consumed, and the surplus of PV production will charge the storage or will be injected into the grid. Vice versa, in case of low production and high loads $E_{PV\_1day\text{-}ahead} < E_{loads\_1day\text{-}ahead,\ 6\ a.m.\text{–}6\ p.m.}$, an SBMS is necessary not only for 1-day head but also for the day after. In this case, the PV production cannot satisfy local loads and storage has to be able to reduce loads for two nights, and the *TDT* will be equal to 36 h.

Batteries are expensive [40], and considering a storage with a too high capacity is not cost-effective for a grid-connected plant. For this reason, in the present work, the BESS can distribute the stored energy in a maximum *TDT* = 36 h. It means that

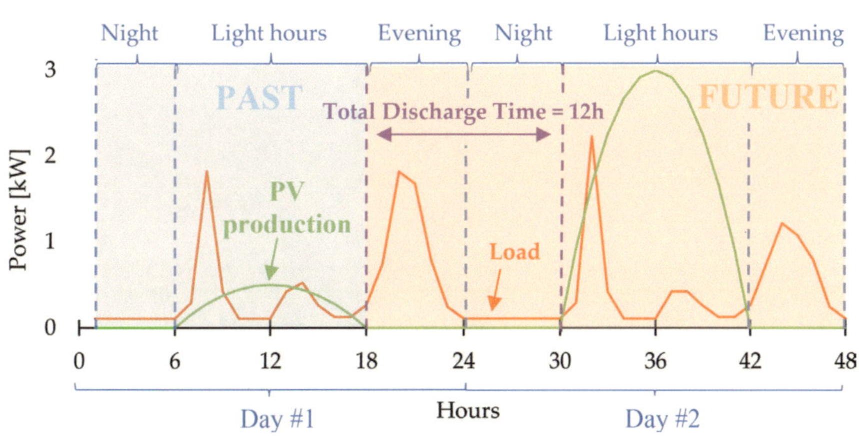


**Figure 5.**
*Example of PV and load profiles for 2 days.*

storage must be able to supply the load when a single cloudy day occurs (2 nights and 1 day).

**Figure 5** shows an example of PV and load profiles for 2 days: in the first day, the PV production is low, while the second one is a clear sky day. At 6:00 p.m. of day #1, the procedure starts with the converter downloading forecast for day #2: supposing a correct forecast, the result is a provisional high PV production. Thus, the BESS will manage the discharge of the storage from the evening of day #1 at 6:00 p.m. to the morning of day #2 at 6:00 a.m. (12 h). After 6:00 a.m. of day #2, storage and loads will again be mainly supplied by the PV production.

The second case is shown in **Figure 6**. It presents an example of PV and load profiles for 3 days: in the first and second days, the real PV production is low, while the third one is a clear sky day. At 6:00 p.m. of day #1, the converter downloads forecast for day #2: supposing a correct forecast, the result is a provisional low PV production. Thus, the BESS will manage the discharge of the storage until the morning of day #3 (a total of 36 h, from 18 to 54 h in **Figure 6**).

### 3.3 Selection of the storage management strategy

After the definition of the total discharge time (*TDT*), the procedure continues with the second part; i.e., the definition of the storage management strategy. The *SOC* is calculated at 6:00 p.m. by the BESS, which uses appropriate models starting from the real-time measurement of voltage and ambient temperature of batteries, as described in Section 2.1. The rated capacity of the storage and the *SOC* permit to calculate the energy that can be provided to the loads $E_{batt,disch}$. The estimated energy production $E_{PV\_1day\text{-}ahead}$ is the same quantity used in the previous step, while the consumption $E_{load}$,*TDT* corresponds to the estimated loads during the *TDT* (**Figure 7**). These raw energy quantities are compared and it is defined if there is an energy deficit $E_{PV\_1day\text{-}ahead} + E_{batt,disch} \geq E_{load,TDT}$ or surplus $E_{PV\_1day\text{-}ahead} + E_{batt,disch} < E_{load,TDT}$.

If the PV production and the storage can satisfy the load $E_{PV\_1day\text{-}ahead} + E_{batt,disch} \geq E_{load,TDT}$ in the selected *TDT*, no advanced management of the batteries is required (BMS Strategy #1).

On the contrary, if loads are too high $E_{PV\_1day\text{-}ahead} + E_{batt,disch} < E_{load,TDT}$, peak shaving strategy (BMS Strategy #2) or appropriate discharge profiles (BMS Strategy #3) are adopted. To select the most appropriate method between BMS

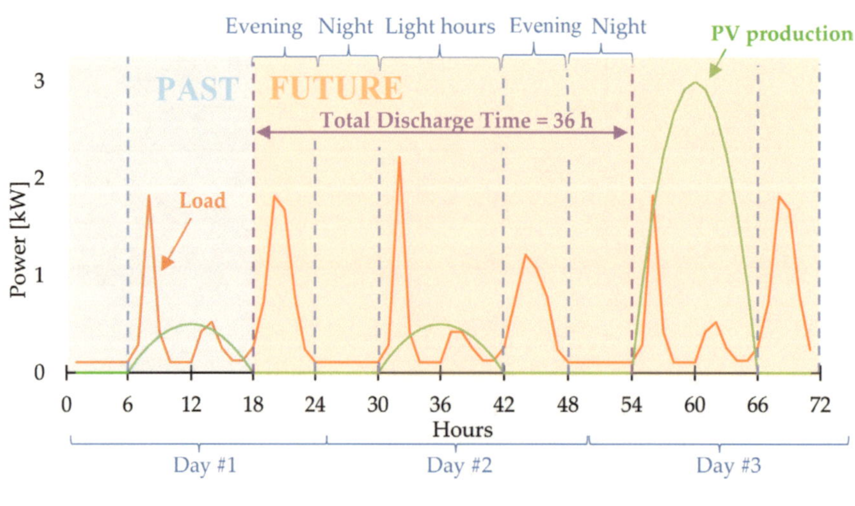


**Figure 6.**
*Example of PV and load profiles for 3 days.*

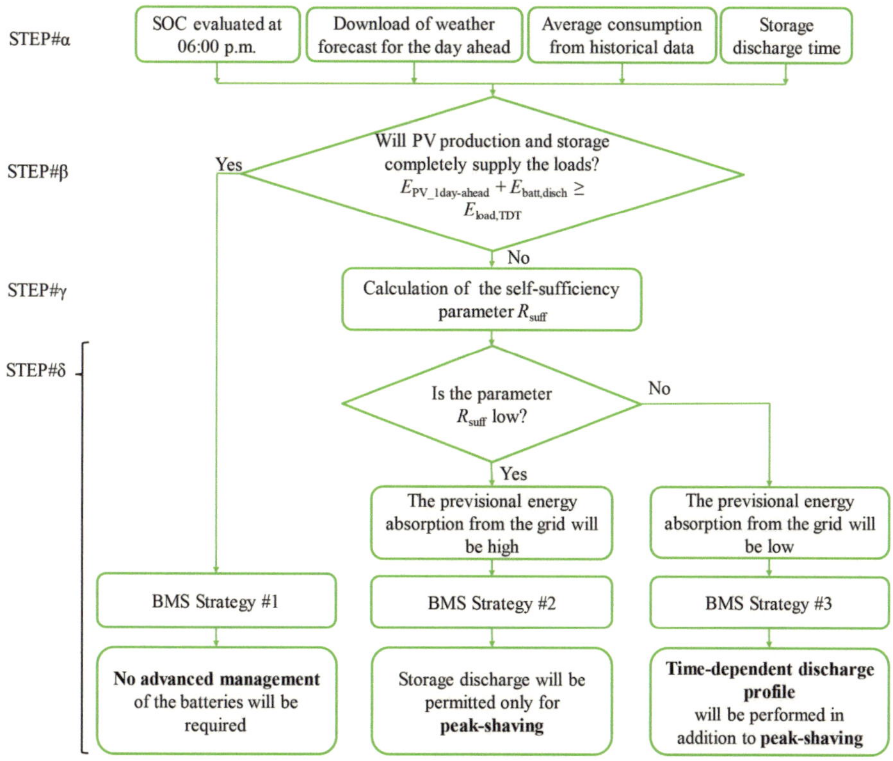


**Figure 7.**
*Definition of storage management strategy.*

Strategy #2 and BMS Strategy #3, a provisional self-sufficiency $R_{suff}$ parameter, that is, the ratio between the provisional PV production plus the available energy from the battery, and the provisional local loads, is calculated:

$$R_{suff} = \frac{E_{PV_{1day}\text{-}ahead} + E_{batt,disch}}{E_{load,TDT}} \tag{7}$$

When the ratio $R_{suff}$ is lower than a user-defined threshold $R_{thres}$, the BMS Strategy #2 is adopted: the local generators and the storage will provide a low quantity of energy to the loads, which will be mainly supplied by the grid. It can result in high absorption peaks. In this case, the low energy quantity stored in the batteries will be used only when loads exceed a maximum limit $P_{load,max}$, such as the contracted power absorption limit or another user-defined threshold. The BMS Strategy #3 is adopted when the ratio $R_{suff}$ is higher than the user-defined threshold $R_{thres}$ and lower than unit value. This case is better than the previous one, because great part of loads will be supplied by PV and storage and the quote from the grid is low.

### 3.4 Implementation of storage management strategies

The storage management strategies consist of peak shaving and of a time-dependent discharge profile. According to the procedure described in the previous subsection, when the storage energy is much lower than loads, only the peak shaving technique is adopted (BMS Strategy #2). Thus, batteries are discharged only when strictly necessary, i.e., when a load peak occurs. In particular, storage will be discharged only by the quota exceeding an user-defined limit $P_{load,max}$.

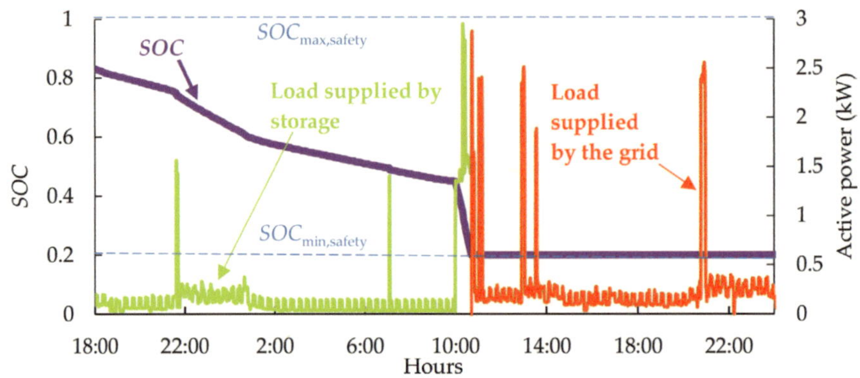


**Figure 8.**
*Example of load and SOC profiles in case of basic BMS.*

In the other case, if the energy stored in the batteries is slightly lower than loads, the charge is used both for baseload supply and peak shaving (BMS Strategy #3). Nevertheless, the exact time schedule of loads is not predictable and it is not possible to know when the load peaks will occur. In the worst case, storage will be discharged soon in the evening, while the peak will be in the next early morning, when batteries are already empty. For this reason, the SBMS limits the discharge of batteries during time with the definition of different levels of minimum $SOC_{min,x}$ for an user-defined number of time slots $x$, in which the *TDT* is divided. According to the procedure proposed in Section 3.2, in case of $TDT = 12$ h, the number of time slots $x = 2$, otherwise with $TDT = 36$ h, the time slots are 5 ($x = 5$). The $SOC_{min,x}$ limits are defined in order to distribute the stored energy proportionally to the provisional energy consumption. Thus, $SOC_{min,x}$ limits are calculated starting from the *SOC* of the storage, measured in real time by the BMS, and the provisional energy consumptions:

$$SOC_{min,slot\,x} = SOC \cdot \left(1 - \frac{E_{load,slot\,x}}{E_{load.TDT}}\right) \quad (8)$$

where $E_{load,\,slot\,x}$ is the provisional energy that will be required by loads in the time slot $x$. For example, let us suppose that the TDT is 12 h and the overall required load will be 10 kWh. In particular, during the evening (from 6:00 p.m. to midnight), the required load will be 4 kWh, and during the next night (from midnight to 6:00 a.m.), the load will be 6 kWh. The stored energy will be discharged as follows: 40% during the evening and 60% during the night. In this example, the storage is considered initially full and with a minimum $SOC_{min,safety} = 0.2$.

**Figure 8** shows an example of load and *SOC* profiles in case of a basic battery management. In this case, the storage is charged when PV production is higher than loads and batteries are empty; on the contrary, storage is discharged if PV production is lower than loads [23]. The only limitation in charge/discharge is performed to avoid fast degradation of batteries, by limiting the *SOC* in a safety range $SOC_{min,safety} < \text{SOC} < SOC_{max,safety}$. For sake of simplicity, it is considered a rainy day and the production from the PV generator is negligible. In case of lithium batteries (**Figure 8**), the minimum level $SOC_{min,safety}$ generally corresponds to $SOC_{safety} \approx 20\%$, while in case of lead-acid batteries, it can reach 50% [41, 42]. In the example of **Figure 8**, the storage supplies the loads until 10:50 a.m., when the $SOC_{min,safety}$ is reached. After that, only the grid supplies the load and the highest absorption peak is not limited ≈2.9 kW.

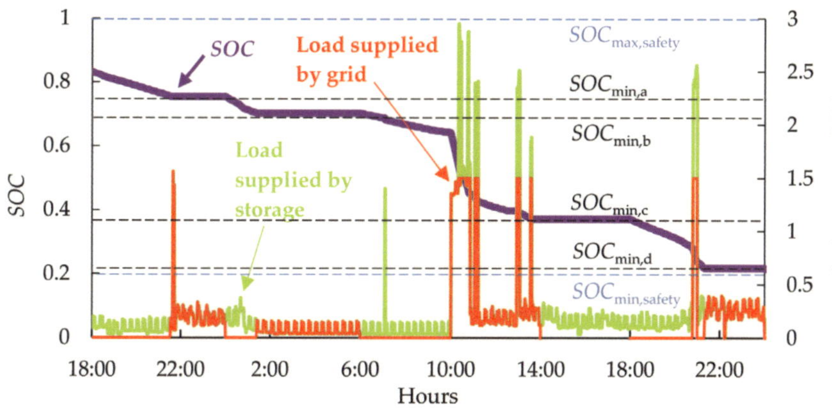


**Figure 9.**
*Example of load and SOC profiles in case of the proposed SBMS.*

**Figure 9** shows an example of load and *SOC* profiles in case of the SBMS, which reduces the absorption peaks from the grid. In this case, the *SOC* cannot drop down under a temporary minimum $SOC_{min,a}$ = 75% before midnight; then, the discharge is limited by $SOC_{min,b}$ = 70% between midnight and 06:00 a.m. Between 06:00 a.m. and 06:00 p.m., the minimum admitted $SOC_{min,c}$ is 37%. Then, between 06:00 p.m. and midnight, the limit $SOC_{min,d}$ = 22%. Finally, after 06:00 p.m., the last limit corresponds to the same level of the basic management $SOC_{safety} \approx 20\%$, which is a typical value to preserve life of lithium batteries. The main difference from the basic management consists of a small reserve in storage, which is always present, and the absorption peaks are always reduced. On the other hand, preserving the storage partially charged could reduce the self-sufficiency. The best solution consists of the abovementioned *SOC* levels selected to reduce absorption peak and keep as high as possible the self-sufficiency level.

## 4. Simulation results

### 4.1 Inputs parameters and constraints of the simulations

Simulations of the PV-storage system are performed for the entire month of December with a 1-min time step for both basic and proposed BMS to compare their performance. During winter, the PV production is low, batteries are often empty, and the development of an efficient BMS is necessary to reduce the absorption peaks from the grid. On the contrary, in summer, PV generation generally charges storage and directly supplies part of the loads.

The optimal management of the storage is investigated in case of different sizes of the PV system $P_{PV,r}$ and different capacities of the battery $C_{bat}$. Regarding $P_{PV,r}$, it ranges between 2 and 6 kW$_p$ with a step of 1 kW, while the storage capacity $C_{bat}$ is in the range 1–5 kWh (step of 1 kWh). The management parameters are the power value $P_{load,max}$ beyond which the peak shaving strategy works and the threshold $R_{thres}$. The power limitation $P_{load,max}$ ranges between 0.5 and 2 kW with a step of 0.5 kW, while the user-defined threshold $R_{thres}$ varies between 50 and 80% (step of 10%). Regarding the loads, the measured consumption profile of a domestic user (a family composed of two persons) located in Northern Italy (45.05° Nord, 7° 40' Est) is used. The annual consumption of the domestic user analyzed in the case study is ≈2800 kWh/year and its loads correspond to typical home appliances (e.g., hairdryer, oven, personal computer, lighting, and electric water heater).

### 4.2 Case study

The results of the simulation show that the proposed BMS decreases the peaks of absorption from the grid with respect to a traditional management. The results are interesting especially in case of a small storage, while in case of higher storage capacity, there are negligible differences between the two managements. **Figure 10** shows case #1: it corresponds to the analysis of 2 days of simulation for a PV system with $P_{PV,r}$ = 4 kW and a storage system with $C_{bat}$ = 2 kWh. In the graphs, in case of battery discharge, the sign of the power supplied by the storage to the loads is negative. The 2 days are characterized by cloudy and rainy conditions, and the PV production is low. The proposed BMS calculates the provisional energy balance and a huge lack in storage is predicted; thus, the peak shaving method is used (BMS Strategy #2). Before 6:00 p.m., all the loads are supplied by PV and storage; then, peak shaving is applied and only the quota exceeding $P_{load,max}$ = 2 kW is satisfied by batteries. The saved energy is then preserved and used to shave loads during the second day, with the result of keeping the absorption from the grid always ≤2 kW.

On the contrary, if a standard BMS is used (**Figure 11**), all the stored energy is consumed before the end of the evening of the first day; furthermore, there is no energy from storage to supply the load peaks during the second day. The result is a maximum absorption peak of ≈4.2 kW: during these days, the proposed SBMS reduces the absorption peak of ≈50%.

**Table 2** shows the energy balance of the case #1 related to **Figures 10** and **11**. With the proposed SBMS, the maximum power absorbed from the grid is half, while the deviations in terms of self-sufficiency and injected energy into the grid are negligible. Nevertheless, there is an increase in grid absorption: to guarantee power for peak shaving, a residual energy is kept in the storage, and at 6:00 p.m. of the second day *SOC*≈0.7.

A second simulation is shown in **Figure 12**. The case #2 is characterized by two different days with respect to case #1: a negligible PV production occurs in both days, while the sizes of PV and storage systems and loads are the same of case #1. The provisional energy balance predicts that the energy in the storage will supply great part of the loads, but it will be not sufficient to supply them totally. The

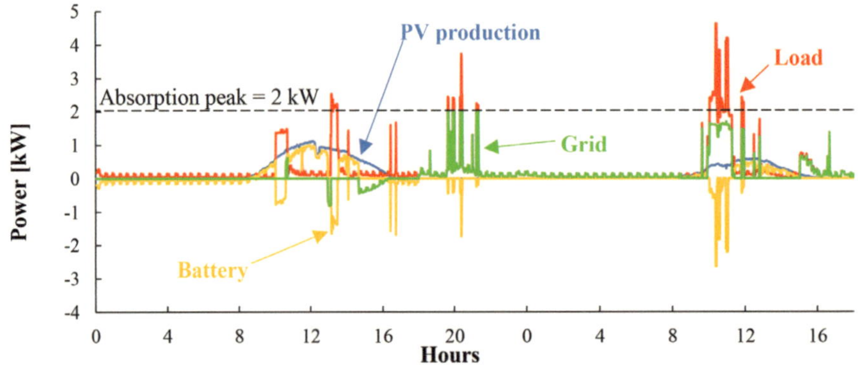


**Figure 10.**
*Power profiles for case #1 with the proposed SBMS.*

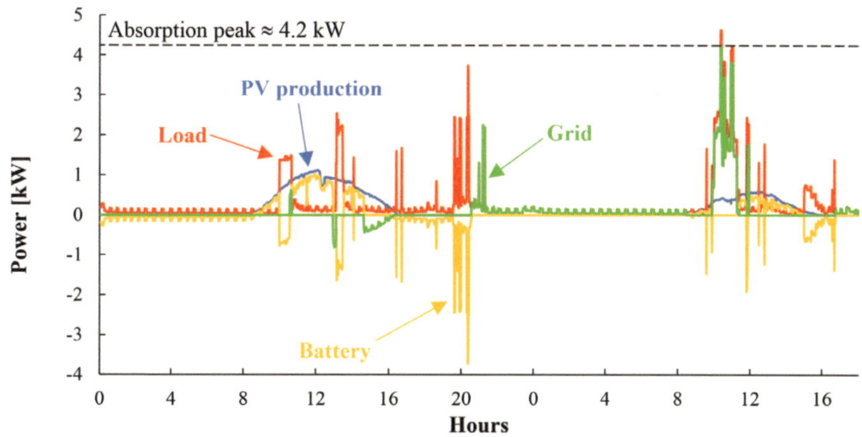


**Figure 11.**
*Power profiles for case #1 with standard BMS.*

provisional self-sufficiency parameter is $R_{suff} > R_{thres}$ (with $R_{thres} = 50\%$); thus, the converter selects the BMS Strategy #3. The most interesting part corresponds to the time window 6:00 a.m.–6:00 p.m. of the second day. Batteries start discharging at 6:00 a.m. and when peaks occur (10:00–12:00 a.m.), only the quota exceeding $P_{load,max} = 2$ kW is satisfied by batteries. In the same way, the other absorption peak occurring at 9:00 p.m. is shaved, thanks to the preserved energy in the storage. The maximum absorption peak is 2.4 kW.

On the contrary, in the same conditions, a traditional BMS would discharge the battery before 10:00 a.m. and the absorption peak would be 2.8 kW (higher than the proposed SBMS of ≈14%).

Finally, in **Table 3**, the above-described combination #A and other three combinations of PV and storage sizes, which permit to obtain significant improvements, are presented. The power and energy results of the proposed SBMS are compared to the standard BMS. In all the other cases, the improvement in terms of maximum absorption from the grid is confirmed, ranging from ~9 to ~10%. Regarding the maximum injection into the grid and the energy quantities, their deviations are negligible. The combination #A shows much better results, confirming that the

| | **Proposed BMS** | **Standard BMS** |
|---|---|---|
| Load (kWh) | 11.45 | 11.45 |
| Self-consumption (kWh) | 3.2 | 3.2 |
| Grid absorption (kWh) | 5.5 | 4.58 |
| Grid injection (kWh) | 0.52 | 0.52 |
| Self-sufficiency/load (%) | 28 | 28 |
| Self-consumption/PV production (%) | 43 | 43 |
| Grid injection/load (%) | 4.5 | 4.5 |
| $P_{load,max}$ (kW) | 2 | 4.22 |
| Injection peak (kW) | −0.81 | −0.81 |

**Table 2.**
*Energy results for case #1.*

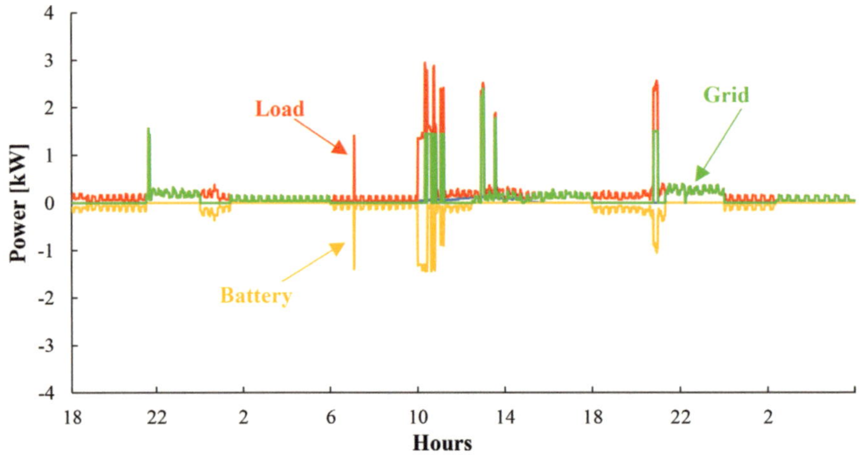


**Figure 12.**
*Power profiles for case #2 with proposed SBMS.*

| Combinations | $P_{PV,r}$ (kW) | $C_{bat}$ (kWh) | $P_{load,max}$ (kW) | $R_{thres}$ (%) | Standard BMS $P_{max,absorbed}$ (kW) | Proposed BMS $P_{max,absorbed}$ (kW) | Improvement (%) |
|---|---|---|---|---|---|---|---|
| #A | 4 | 2 | 2 | 50 | 4.22 | 2.56 | 39 |
| #B | 2 | 1 | 2 | 70 | 4.42 | 4 | 9.5 |
| #C | 3 | 2 | 2 | 80 | 4.33 | 3.91 | 9.7 |
| #D | 5 | 5 | 1 | 60 | 4.10 | 3.74 | 8.8 |

**Table 3.**
*Results of the alternative configurations.*

performance of the proposed SBMS increases when the PV system size is high and when the storage is undersized. In addition, a low value of $R_{suff}$ permits to increase the use of peak shaving, without affecting the energy balance.

## 5. Conclusions

In the present work, smart BMS for residential users with a grid-connected PV-storage system is proposed. The BMS is Internet-connected and it downloads 1-day ahead weather forecasts, which are used to obtain a provisional energy production for the PV generator. These data are compared with load estimations, based on historical data. The result is a provisional energy balance, which is used by the BMS to select the best strategy to discharge batteries. In particular, the BMS preserves battery charge, when high load and low production is expected, and performs peak shaving, when loads exceed a user-defined limit. The combination of these methods results in a reduction in absorption peaks from the grid, with negligible variations in terms of self-sufficiency. The proposed BMS is efficient in case of undersized batteries, where the energy available in the storage is often not sufficient to supply all the loads. For example, in case of a family composed of two persons with a PV plant with rated power 4 kW and a storage of 2 kWh, the reduction in absorption peak from the grid during winter days varies from 39 to 50%. Other combinations of PV and storage sizes are investigated and improvements in terms of peaks reduction are generally around 10%.

## Nomenclature

### Acronyms

| | |
|---|---|
| ANN | artificial neural networks |
| BS | battery system |
| BMS | battery management system |
| BESS | battery energy storage system |
| MPPT | maximum power point tracker |
| PV | photovoltaic |
| PVGIS | Photovoltaic Geographical Information System |
| RES | renewable energy sources |
| SBMS | smart battery management system |
| STC | standard test conditions |

### Symbols

| | |
|---|---|
| $\gamma_{th}$ | temperature factor of power of PV generator (%/°C) |
| $\eta_{cabl}$ | Joule losses |
| $\eta_{charge}$ | charge efficiency of the battery |
| $\eta_{DC/AC}$ | DC/AC conversion losses |
| $\eta_{dirt}$ | losses due to dirt |
| $\eta_{mis}$ | losses due to mismatch |
| $\eta_{mix}$ | global losses of PV generator |
| $\eta_{MPPT}$ | DC/DC conversion losses |
| $\eta_{refl}$ | losses due to reflection |
| $\eta_{shad}$ | losses due to shadings |
| $\eta_{therm}$ | thermal losses of PV generator |
| $C_{bat}$ | rated capacity of the battery (kWh) |
| $E_{batt,disch}$ | battery energy provided to the loads (kWh) |
| $E_{load}$ | energy consumptions (kWh) |
| $E_{loads\_1day\text{-}ahead,\ 6\ a.m.–6\ p.m.}$ | 1-day ahead expected loads in the time slot 6 a.m.–6 p.m. (kWh) |
| $E_{load,\ slot\_x}$ | provisional loads in the time slot x (kWh) |
| $E_{load,TDT}$ | estimated loads during the *TDT* (kWh) |
| $E_{PV}$ | PV production (kWh) |
| $E_{PV\_1day\text{-}ahead}$ | 1-day ahead expected PV production (kWh) |
| $G$ | solar irradiance (W/m$^2$) |
| $G_{NOCT}$ | solar Irradiance at NOCT conditions (W/m$^2$) |
| $NOCT$ | nominal operating cell temperature (°C) |
| $P_{AC}$ | AC power production of PV generator (kW) |
| $P_{bat}$ | power exchanged by the battery (kW) |
| $P_{load,max}$ | maximum load power satisfied by the grid in case of peak shaving strategy (kW) |
| $P_{max,absorbed}$ | maximum power absorbed from the grid (kW) |
| $P_{PV,r}$ | rated power of PV generator (kW) |
| $R_{suff}$ | provisional self-sufficiency parameter |
| $R_{thre}$ | threshold for the parameter $R_{suff}$ |
| $SOC$ | state of charge of the battery |
| $SOC_{max,safety}$ | maximum safety limit of SOC |
| $SOC_{min,a}$ | minimum SOC in the time slot a |
| $SOC_{min,b}$ | minimum SOC in the time slot b |

| | |
|---|---|
| $SOC_{min,c}$ | minimum SOC in the time slot c |
| $SOC_{min,safety}$ | minimum safety value of SOC |
| $SOC_{min,slot\_x}$ | minimum SOC in the time slot x |
| $SOH$ | state of health of the battery |
| $\Delta t$ | simulation time step (min) |
| $t$ | simulation time (min) |
| $T_a$ | ambient temperature (°C) |
| $T_c$ | temperature of PV cells (°C) |
| $T_{STC}$ | temperature at standard test conditions (°C) |
| $TDT$ | total discharge time of the battery (h) |
| $x$ | user-defined number of time slots in which the TDT is divided |

## Author details

Filippo Spertino*, Alessandro Ciocia, Paolo Di Leo, Gabriele Malgaroli and Angela Russo
Energy Department "Galileo Ferraris", Politecnico di Torino, Italy

*Address all correspondence to: filippo.spertino@polito.it

## References

[1] Lahon R, Gupta CP. Energy management of cooperative microgrids with high-penetration renewables. IET Renewable Power Generation. 2018;**12**(6):680-690. DOI: 10.1049/iet-rpg.2017.0578

[2] Rubino S, Mazza A, Chicco G, Pastorelli M. Advanced control of inverter-interfaced generation behaving as a virtual synchronous generator. In: Proceedings of the 2015 IEEE Eindhoven PowerTech; Eindhoven; 2015. pp. 1-6

[3] Chicco G, Mazza A. An overview of the probability-based methods for optimal electrical distribution system reconfiguration. In: Proceedings of the 2013 4th International Symposium on Electrical and Electronics Engineering (ISEEE); Galati; 2013. pp. 1-10

[4] Mazza A, Chicco G, Andrei H, Rubino M. Determination of the relevant periods for intraday distribution system minimum loss reconfiguration. International Transactions on Electrical Energy Systems. 2015;**25**:1992-2023. DOI: 10.1002/etep.1941

[5] Narducci G, Mazza A, Chicco G, Bompard E. Battery storage application for covering the mismatch between scheduled and CHP plant production. In: Proceedings of the 2017 52nd International Universities Power Engineering Conference (UPEC); Heraklion; 2017. pp. 1-6

[6] Spertino F, Ahmad J, Chicco G, Ciocia A, Di Leo P. Matching between electric generation and load: Hybrid PV-wind system and tertiary-sector users. In: Proceedings of the 50th International Universities Power Engineering Conference (UPEC); Trent; 2015. pp. 1-6

[7] Ciocia A, Chicco G, Di Leo P, Gai M, Mazza A, Spertino F, Hadj-Said N. Voltage control in low voltage grids: A comparison between the use of distributed photovoltaic converters or centralized devices. In: Proceedings of the IEEE International Conference on Environment and Electrical Engineering and IEEE Industrial and Commercial Power Systems Europe (EEEIC/I&CPS Europe); Milan; 2017. pp. 1-6

[8] Akarslan E, Hocaoglu FO. Electricity demand forecasting of a micro grid using ANN. In: Proceedings of the 9th International Renewable Energy Congress (IREC); Hammamet; 2018. pp. 1-5

[9] Koller M, Borsche T, Ulbig A, Andersson G. Defining a degradation cost function for optimal control of a battery energy storage system. In: Proceedings of the 2013 IEEE Grenoble Conference; Grenoble; 2013. pp. 1-6

[10] Bizzarri F, Bongiorno M, Brambilla A, Gruosso G, Gajani GS. Model of photovoltaic power plants for performance analysis and production forecast. IEEE Transactions on Sustainable Energy. 2012;**4**:278-285. DOI: 10.1109/TSTE.2012.2219563

[11] Brenna M, Foiadelli F, Longo M, Zaninelli D. Mint: Solar radiation and load power consumption forecasting using neural network. In: Proceedings of the 2017 6th International Conference on Clean Electrical Power (ICCEP); Santa Margherita Ligure; 2017. pp. 726-731

[12] Lin Q, Yin M, Shi D, Qu H, Huo J. Optimal control of battery energy storage system integrated in PV station considering peak shaving. In: Proceedings of the 2017 Chinese Automation Congress (CAC); Jinan; 2017. pp. 2750-2754

[13] Wang B, Zarghami M, Vaziri M. Energy management and peak-shaving in grid-connected photovoltaic systems integrated with battery storage. In: Proceedings of the 2016 North American Power Symposium (NAPS); Denver; 2016. pp. 1-5

[14] Pholboon S, Sumner M, Christopher E, Norman SA. Real-time battery management algorithm for peak demand shaving in small energy communities. In: Proceedings of IEEE Innovative Smart Grid Technologies Latin America (ISGT LATAM); Montevideo; 2015. pp. 19-24

[15] Ranaweera I, Midtgård OM, Korpås M, Farahmand H. Control strategies for residential battery energy storage systems coupled with PV systems. In: Proceedings of the 2017 IEEE International Conference on Environment and Electrical Engineering and 2017 IEEE Industrial and Commercial Power Systems Europe (EEEIC); Milan; 2017. pp. 1-6

[16] Marinopoulos A, Papandrea F, Reza M, Norrga S, Spertino F, Napoli R. Grid integration aspects of large solar PV installations: LVRT capability and reactive power/voltage support requirements. In: Proceedings of the 2011 IEEE Trondheim PowerTech; Trondheim; 2011. pp. 1-8

[17] Lamberti F, Calderaro V, Galdi V, Piccolo A, Gross G. Long-term performance in providing voltage support by PV and storage systems in distribution networks. In: Proceedings of the 2016 IEEE Power and Energy Society General Meeting (PESGM); Boston, MA; 2016. pp. 1-5

[18] Del Giudice A, Wills A, Mears A. Development of a planning tool for network ancillary services using customer-owned solar and battery storage. In: Proceedings of the 2017 IEEE PES Innovative Smart Grid Technologies Conference Europe (ISGT); Torino; 2017. pp. 1-6

[19] Praiselin WJ, Edward JB. Improvement of power quality with integration of solar PV and battery storage system based micro grid operation. In: Proceedings of the 2017 Innovations in Power and Advanced Computing Technologies (i-PACT); Vellore; 2017. pp. 1-5

[20] Aachiq M, Oozeki T, Iwafune Y, Fonseca JGS Jr. Reduction of PV reverse power flow through the usage of EV's battery with consideration of the demand and solar radiation forecast. In: Proceedings of 2013 IEEE International Electric Vehicle Conference (IEVC); Santa Clara, CA; 2013. pp. 1-3

[21] Ahmad J, Spertino F, Ciocia A, Di Leo P. A maximum power point tracker for module integrated PV systems under rapidly changing irradiance conditions. In: Proceedings of the 2015 IEEE 5th Conference on Consumer Electronics (ICCE); Berlin; 2015. pp. 519-520

[22] European Commission. Joint Research Centre, Photovoltaic Geographical Information System [Internet]. Available from: http://re.jrc.ec.europa.eu/pvgis/ [Accessed: June 06, 2018]

[23] Spertino F, Ciocia A, Cocina V, Di Leo P. Renewable sources with storage for cost-effective solutions to supply commercial loads. In: Proceedings of the IEEE International Symposium on Power Electronics Electrical Drives Automation and Motion (SPEEDAM); Anacapri, Italy; 2016. pp. 242-247

[24] Whitaker CM, Townsend TU, Wenger HJ, Iliceto A, Chimento G, Paletta F. Effects of irradiance and other factors on PV temperature coefficients. In: Proceedings of the Conference Record of the Twenty-Second

IEEE Photovoltaic Specialists Conference—1991; Vol. 1. Las Vegas, NV; 1991. pp. 608-613

[25] Smith RM, Jordan DC, Kurtz SR. Outdoor PV module degradation of current-voltage parameters. In: World Renewable Energy Forum; May 13-17 2012

[26] Muller M, Marion B, Rodriguez J. Evaluating the IEC 61215 Ed.3 NMOT procedure against the existing NOCT procedure with PV modules in a side-by-side configuration. In: Proceedings of the 2012 38th IEEE Photovoltaic Specialists Conference; Austin, TX; 2012. pp. 697-702

[27] Spertino F, Ahmad J, Ciocia A, Di Leo P. Techniques and experimental results for performance analysis of photovoltaic modules installed in buildings. Energy Procedia. 2017;**111**:944-953. DOI: 10.1016/j.egypro.2017.03.257

[28] Khanum KK, Rao A, Balaji NC, Mani M, Ramamurthy PC. Performance evaluation for PV systems to synergistic influences of dust, wind and panel temperatures: Spectral insight. In: Proceedings of the 2016 IEEE 43rd Photovoltaic Specialists Conference (PVSC); Portland, OR; 2016. pp. 1715-1718

[29] Tanesab J, Parlevliet D, Whale J, Urmee T, Pryor T. The contribution of dust to performance degradation of PV modules in a temperate climate zone. Solar Energy. 2015;**120**:147-157. DOI: 10.1016/j.solener.2015.06.052

[30] Spertino F, Ciocia A, Di Leo P, Tommasini R, Berardone I, Corrado M, Infuso A, Paggi M. A power and energy procedure in operating photovoltaic systems to quantify the losses according to the causes. Solar Energy. 2015;**118**:313-326. DOI: 10.1016/j.solener.2015.05.033

[31] Spertino F, Ahmad J, Di Leo P, Ciocia A. A method for obtaining the I-V curve of photovoltaic arrays from module voltages and its applications for MPP tracking. Solar Energy. 2016;**139**:489-505. DOI: 10.1016/j.solener.2016.10.013

[32] Hong Y, Yoo T, Chac K, Back K, Kim YS. Efficient maximum power point tracking for a distributed system under rapidly changing environmental conditions. IEEE Transactions on Power Electronics. 2015;**30**:4209-4218. DOI: 10.1109/TPEL.2014.2352314

[33] Spertino F, Ciocia A, Corona F, Di Leo P, Papandrea F. An experimental procedure to check the performance degradation on-site in grid-connected photovoltaic systems. In: Proceedings of the 2014 IEEE 40th Photovoltaic Specialist Conference (PVSC); Denver, CO; 2014. pp. 2600-2604

[34] Waag W, Sauer DU. Secondary Batteries – Lead–Acid Systems | State-of-Charge/Health. In: Garche J. editor. Encyclopedia of Electrochemical Power Sources. 1th ed; 2009:793-804. DOI: 10.1016/B978-044452745-5.00149-0

[35] Einhorn M, Conte V, Kral C, Fleig J. Comparison of electrical battery models using a numerically optimized parameterization method. In: Proceedings of the IEEE Vehicle Power and Propulsion Conference; Chicago, USA; 2011

[36] Hussein AH, Batarseh I. An overview of generic battery models. In: Proceedings of the IEEE Power and Energy Society General Meeting; San Diego, USA; 2011

[37] Chen M, Rincon-Mora GA. Accurate electrical battery model capable of predicting runtime and I-V performance. IEEE Transactions on Energy Conversion. 2006;**21**:504-511. DOI: 10.1109/TEC.2006.874229

[38] Einhorn M, Conte V, Kral C, Fleig J. Comparison of electrical battery models using a numerically optimized parameterization method. In: Proceedings of the 2011 IEEE Vehicle Power and Propulsion Conference; Chicago, USA; 2011. pp. 1-7

[39] Cacciato M, Nobile G, Scarcella G, Scelba G. Real-time model-based estimation of SOC and SOH for energy storage systems. IEEE Transactions on Power Electronics. 2017;**32**:794-803. DOI: 10.1109/TPEL.2016.2535321

[40] Tang DH, Eghbal D. Cost optimization of battery energy storage system size and cycling with residential solar photovoltaic. In: Proceedings of the 2017 Australasian Universities Power Engineering Conference (AUPEC); Melbourne, VIC; 2017. pp. 1-6

[41] Anuphappharadorn S, Sukchai S, Sirisamphanwong C, Ketjoy N. Comparison the economic analysis of the battery between lithium-ion and lead-acid in PV stand-alone application. Energy Procedia. 2014;**56**:352-358. DOI: 10.1016/j.egypro.2014.07.167

[42] Percy SD, Aldeen M, Rowe CN, Berry A. A comparison between capacity, cost and degradation in Australian residential battery systems. In: Proceedings of the 2016 IEEE Innovative Smart Grid Technologies; Melbourne; 2016

Chapter 5

# Integration of Advanced Technologies for Efficient Operation of Smart Grids

*Radu Porumb and George Serițan*

## Abstract

The current transition of electrical power systems toward smart grids is encompassing a fundamental change in their structure, as well as operation. This is setting the path to be followed by the hardware and software embedded in electrical power systems, as well as technology adaptation to the "open-source" customers' needs and consumption patterns. This chapter is following the evolution of energy sector, accompanied by constant improvements of technology, which is providing increasingly complex hardware, which embeds power quality improvement devices, for an efficient operation of electrical power assets. This chapter presents a comprehensive survey of continuous advances of renewable energy sources and storage technology which have started the transformation of end users into energy-efficient and clean prosumers, underlining the subsequent energy markets support of peer-to-peer energy trading through novel technologies as blockchain.

**Keywords:** advanced technology, blockchain, energy efficiency, micro-grid, power quality, smart grid

## 1. Introduction

At present, the general trend of evolution of society has as a catalyst, the technological revolution. Through the continuous development of information analysis and synthesis capabilities through cloud computing technologies, manufacturing capabilities through 3D printing, as well as local power generation and storage capabilities, the industry is experiencing a huge leap across multiple plans.

Also, at the social level, the need to improve the state of the environment, as a *sine qua non* condition of increasing the quality of life, is felt more and more. This pressure has generated mechanisms to control the sources of pollution, with the energy sector being one of the main actors in this respect. Thus, it can be said that energy efficiency is a vital resource for environmental protection.

Energy efficiency has gained a major significance in the design and implementation of national and transnational projects, due to the importance of concepts and measures resulting from the specific analysis of energy processes. From the perspective of environmental protection, the analogy that each consumed kWh leads to a reduction in carbon footprint by about 1 kg of $CO_2$ is essential. Also, the concept of "negajouls" is increasingly being met as the measure of increasing energy efficiency defined by measuring unconsumed energy.

These requirements are driving next-generation engineers toward seamless communication with physical systems with the help of artificial intelligence, turning them into dialog enablers between future smart society and its fundamentally supporting technology.

Ensuring the power quality requires an entire investigation chain, from production, transport, distribution to the public network, and distribution to the user. The transfer of energy through this transformation chain cannot provide an ideal quality of the supplied energy. In this respect, the user, on the basis of careful technical and economic analysis, must accept the quality of the electricity provided by the public supply system (within acceptable risk limits) or adopt measures that require investment to achieve an accepted power quality level for the processes which take place at the end of production chain.

## 2. Energy efficiency

Energy management aims to ensure the judicious and efficient use of energy in order to maximize profit by minimizing energy costs, thus increasing the competitiveness on the market.

Data obtained in the electronic document management system (EDMS) energy management process, although it does not directly lead to energy savings, has a key role to play in identifying savings potential and adopting the most appropriate measures to increase energy efficiency. EDMS is a software program that manages the creation, storage, and control of documents electronically, managing the electronic information within an organization workflow. The implementation of the EDMS allows knowledge of the energy used in each process and provides the premises for achieving an annual energy savings of up to 10% [1–5].

Energy costs are an important element in the cost structure of most products, resulting from production processes. Reducing the energy used ultimately leads to lower production costs and, implicitly, to increased product competitiveness.

Achieving the goals of energy management requires:

- Increasing energy efficiency and reducing energy needs in order to reduce costs without reducing production levels

- Achieving good communication between compartments, specific energy issues, and their accountability regarding energy management

- The development and ongoing use of an energy monitoring system used, the reporting of these values, and the development of specific energy optimization strategies used

- Finding the best ways to increase money savings resulting from investments in energy efficiency of specific production processes by applying the best available technology (BAT) solutions known worldwide

- Developing the interest of all employees in the efficient use of energy and educating them through specific awareness programs to reduce energy losses

- Ensuring the safety of power supply to energy installations

- An overall approach to monitoring and reducing energy needs in any type of organization

- Reducing costs and reducing greenhouse gas emissions

Energy management uses engineering and economic principles to control the energy costs used to provide the necessary services in buildings and industry. Energy cost reductions can come from energy efficiency improvements, and savings can come from changing traditional energy sources with more efficient sources. In this sense, the energy policy implemented within the unit is of particular importance. Five categories of energy policies are met:

- No explicit policy (no delegation of responsibilities on the energy side and no investment to increase energy efficiency)

- Unformulated action directions (occasional tasks and only low-cost measures)

- Incoherent policy (assigning tasks but without responsibilities and only short-term recovery)

- Policy assumed but unmatched by top management (clear allocation of tasks and budget with the same implementation regime as other investments)

- Active involvement of top management (integrating with other forms of management within the unit with the advantage of investing in energy efficiency and reducing carbon footprint)

The main benefits for operators implementing the principles of energy management are:

- Improving the quality of environmental factors

- Improving economic competitiveness by reducing production costs and reducing the intensity of energy used

- Improving energy security

- Reducing the risk of accidents by identifying the weaknesses of the processes

- Improved advertising and image

### 2.1 Energy management standard

In 2011, International Standardization Organization (ISO) has developed the internationally accepted energy management system (EnMS) standard [1]. Based on this, an organization can develop and implement its own energy policy that sets goals, tasks, and action plans that address legal requirements and existing information relative to significant energy uses.

The standard has a principle "plan-realization-control-action (plan-do-check-act (PDCA))" (**Figure 1**).

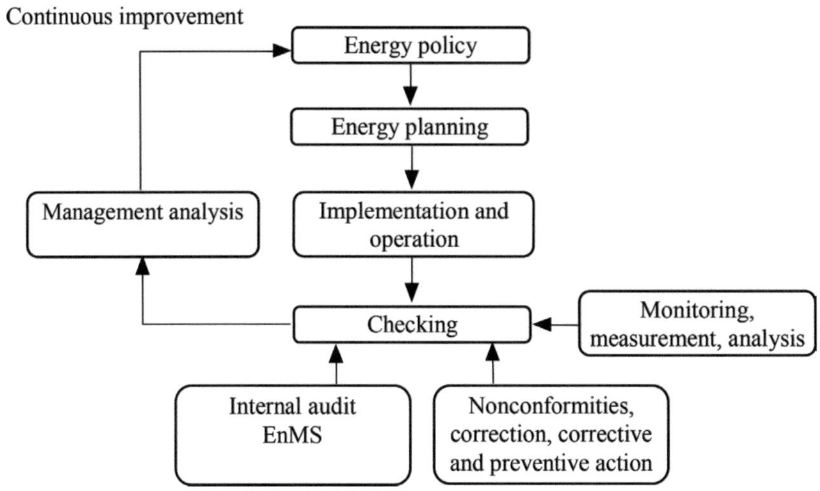


**Figure 1.**
*Energy management system.*

Verification includes process monitoring and measurement and the underlying characteristics of operations to determine energy performance against energy policy and its objectives [2, 6–11]. It also includes reporting on the results obtained during the verification.

Action concerns the adoption of measures to continue to perform as planned within the EnMS.

The activity considered is based on an online and offline measurement and processing system. Energy management practically integrates technical issues, legal/regulatory issues and managerial aspects across the entire energy contour and for all energy operators.

The objective is to operate the system under maximum safety and economic efficiency and ensure the security of electricity supply. Technical issues consist of identifying disturbances that cause deviations from normal operating parameters.

The legal/regulatory aspects consist of establishing the framework for the evaluation and monitoring processes, setting the standards and indicators that characterize the functioning of the system being analyzed and setting the reference levels for the parameters to be monitored. Managerial aspects consist of developing strategies/policies to monitor operating parameters that ensure clear and effective determination of responsibility for deviations from standard regulatory levels and thus achieve performance targets.

Strategies/policies are addressed both to energy operators and to electrical/electrical receiver manufacturers.

An important component of the management of implementation and monitoring of the operating parameters of the analyzed system is the strategies/policies of training and education of the employees.

In general, the management of any activity/project has multiple approaches:

- Approaches from simple to complex. Start with case studies, disseminate information among specialists, then go through pilot project.

- Bottom-up approaches, starting from the technical aspects and phenomenology and reaching the management of the decision-making activity, with aspects of financing, implementation, monitoring, and corrective measures. There is bottom-up pressure on the management to get the decision to implement the project. Typically, these projects have longer implementation times.

- Top-down approaches, based on the information from the literature and its own experience (case studies, research), the management consults/studies and develops decisions and strategies for implementing the project, hierarchically applied from top to bottom. These projects have shorter implementation times.

- Mixed approaches. This approach is often the most effective way to achieve the project implementation goal.

Improving the energy management of a company, based on the analysis of the current management and the experience of the management team, ensures the continuity of the process and the achievement of some performing actions in the field [12–14].

### 2.2 Management projects

Management projects must fit into the enterprise's asset, production, and ancillary activity policy and require a clear and concise definition of the benefits of the company (distribution operator, user, equipment manufacturer, energy service provider). From an economic point of view, benefits have to be quantified in money.

It is worth noting the triangle that should be considered permanently during implementation: cost, time, and benefits. Keeping an optimal balance between these three elements contributes to the success of the project. Regardless of the approach, energy management requires the clear definition of goals, objectives, strategies, policies, and implementation plans with deadlines and responsibilities, as part of the plan-do-check-act (PDCA) of the legislation and regulations under consideration.

Management involves the implementation of the PDCA using the possibilities offered by the development of computer systems and control/control systems based on electronic circuits. Also, the electronic control/control circuits have a significant impact on energy efficiency [15].

When implementing energy management projects where control systems based on power electronics are widely used, both benefits and a number of negative effects can occur. The benefits are primarily determined by accurate measurement of energy performance through the improvement of methods and especially of measuring instruments through the development of information technology [16]. In this way, the conditions were created for the monitoring of the electrical process voltages (electrical voltages and currents) over a long period of time and for the accuracy of measurements.

It was possible to track real-time (online) and offline analysis of the analyzed phenomena, and it was possible to obtain the data necessary for carrying out the forecasting activities.

The experience resulting from the analysis of many energy management programs implemented in different sectors of activity has shown that:

- There can be achieved energy and money savings of 5, 12, or 15%, in very short time, with minimal costs or even without costs, only by applying aggressive energy management.

- Energy and money savings of up to 30% can be achieved, with low and average costs, with a short depreciation period; the application of such measures is frequent.

- By realizing high-cost investments in modern technologies and equipment, savings of 50–70% can be achieved, with depreciation periods of up to 5–6 years.

The main barriers to the promotion of energy-efficient products of a technical, economic, financial, and managerial nature are shown in **Figure 2**.

Although some of the barriers shown in **Figure 2** are objective in nature, many are due to insufficient knowledge of the problem, insufficient awareness, and implications for increasing energy efficiency, especially through the reduction of environmental pollution. New technologies or methods for increasing energy efficiency are not immediately adopted, in most cases, by cost-benefit analysis. An in-depth study of energy savings and product quality enhancements can provide arguments for implementing new technologies and gaining economic benefits.

An important barrier lies in personal and institutional inertia, especially in smaller companies. Also, market conditions and unconvincing marketing can be barriers to making products with new features that incorporate less energy.

An important role is the package of regulations that must ensure the promotion of new energy-efficient technologies. The complexity of the integration process associated with interruption during assembly, commissioning, and calibration results in operators being detained, especially in enterprises that require high production reliability, and who are reluctant to assume the risks of using new technologies. Compliance with the new regulations on the conditions of the use of new technologies, especially those relating to environmental compliance and security conditions, leads to delays in adopting them. When deciding on the implementation of new technologies, the investments needed to rehabilitate an installation and adopt new technology with new production capacity must be compared. In some cases, the relatively long recovery of investment in new technologies also limits decision-making on their implementation. Unlike large users, individual users or

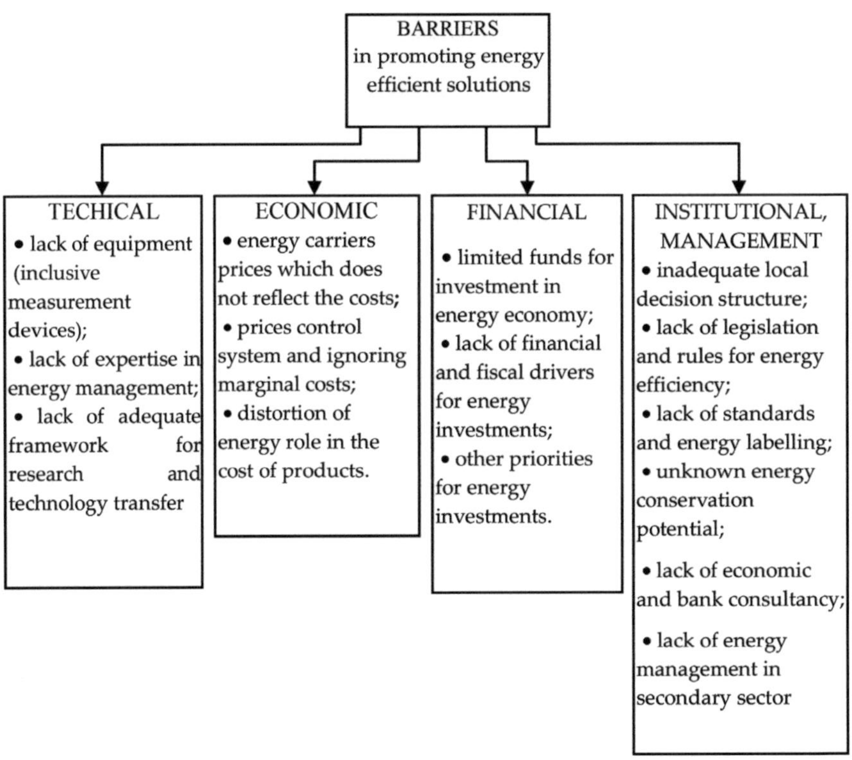


**Figure 2.**
*Barriers in energy-efficient solutions promoting.*

small and medium businesses do not have information, technical and managerial capabilities, and data on efficient funding schemes to conduct a detailed analysis, which limits their access to new technologies. Also, there may be difficulties in putting them into operation and using these technologies.

## 3. Power quality management in smart grids

Monitoring the quality of electrical energy in a node of the electrical network is intended to determine the characteristics of the voltage and current curves as well as the variations in the frequency of the voltage in the network in relation to a set of standardized technical indicators. The monitoring of the quality indicators of the electricity and the adoption of the measures necessary to maintain them at the level stipulated by the quality standards fall within the obligation of the network operators.

Wide implementation of renewable energy sources can have a negative impact on the quality of electricity (voltage variations, harmonic disturbances, voltage fluctuations, and increase in voltage gaps).

The development of smart grids and micro-grids leads to increased electricity production in low-voltage (low-power generation) networks and changing user type (intelligent user systems, electric vehicle charging stations, etc.) which can lead to the emergence of important electromagnetic disturbances that could lead to lowering the level of electricity quality in the nodes of this network.

The implementation of smart grids requires measures adopted at the level of each operator in the power system to ensure the required objectives [11, 17–22].

At the user's level (prosumer):

- Efficient energy use systems
- Energy production from local sources
- Intelligent buildings
- Automation of user equipment

At generation level:

- Adaptive production with a focus on renewable energy sources
- Environmental pollution control for conventional sources

At network level:

- Station automation (SA—substation automation)
- Power quality (PQ) and network monitoring (PM—power monitoring)
- Power management in the system (EMS—energy management system)
- Wide use of power electronics
- Asset management in the system and their monitoring

- Automation of distribution
- Management of distribution systems
- Advanced metering infrastructure (AMI)

At the level of communication systems:

- Ensuring the security of the communication circuits
- The development of communication platforms

The main issues to be monitored in smart grids to ensure the required quality of electricity provided to end users are:

- The direction of power flow in the transmission and distribution networks
- Voltage level in all nodes of the power grid (in classical systems it is sufficient to monitor the voltages in the representative nodes)
- The value of the power factor in all the nodes of the electric network
- System unbalance (especially in low-voltage networks) and neutral conductor loading
- The voltage curve distortion level in all nodes of the electrical network (in classical systems, monitoring at the common point of coupling of the users is sufficient)
- Micro-grid insularity

The data obtained from the monitoring shall transmit information for the adoption of measures to frame the voltages of the nodes of the electrical network within the accepted limits of the quality of the electric power [15]:

- Self-configuring the network to ensure continuity in the feed;
- Ensuring robustness against physical and computer attacks;
- Correlation of the sources of generation with the possibilities of storage of electricity;
- Reactive power control systems;
- Control of voltage curve distortion;
- Control of unbalanced load-limiting systems

## 4. Micro-grid

Intelligent "power" grids are characterized by the widespread use of modern technologies, IT procedures for system management, and extended automation and the use of current energy management systems, a high-quality power supply service by the possibility of self-correcting the effects of defects. Within "smart" electricity networks, a particular attention is paid to communication systems that need to provide data and order transmission for optimal network operation with high energy efficiency and minimum energy prices for users.

Although under the electricity market, the minimization of energy losses in electrical networks is not a preoccupation with the need to ensure the requirements of users and producers; the existence of detailed information on the state of the system can provide the conditions for choosing an energy transmission structure so as to ensure the minimization of energy losses in the conditions of the restrictions determined by the specific requirements of the electricity market.

The operation of renewable energy sources with uncontrollable generated power and the wider availability of micro-networks using local energy sources determine that two-way power circuits can be drawn on the lines of the power system to be taken into account in the determination programs of losses.

Micro-grid development plays an important role in reducing energy losses by limiting the power flow on power lines, while the micro-network is powered by its own sources (wind and solar). The operation of a micro-network with the optimal use of local power sources requires a reliable communication network to acquire all the information required for micro-network management [22–28].

As an example of communication circuits, **Figure 3** shows a micro-circuit diagram with circuits that include information on operating parameters, purchased energy prices, and power source supervisory control to provide users with high quality energy, and reduced losses.

In principle, micro-processing management, based on information on the state of internal sources and the price of electricity on the energy market, must provide the users of the micro-network with the lowest electricity offered. In this respect, it

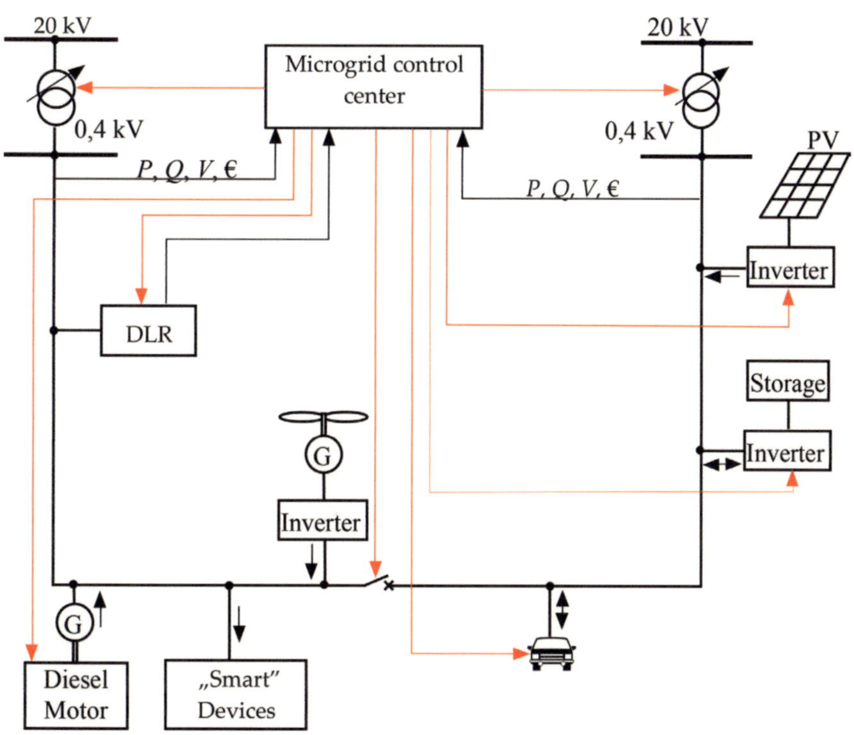


**Figure 3.**
*Energy and information transfer into a micro-grid.*

can use the additional possibilities offered by the "smart" receivers whose load can be changed without affecting the user's "comfort," the energy stored in the battery of the electric cars in the area, as well as the local electrical storage facilities.

## 5. Smart grids and blockchain

The blockchain-based shift in paradigm is brought by a 2008 article of Satoshi Nakamoto [29], which sparked a true revolution for the future use of Internet. The base layer of these changes signifies that it brings fundamental change in the fabric of the Internet, changing it form an *Internet of information* to an *Internet of value* [30]. The core function of the latter is that it enables peers to exchange values over the net, without fear of getting ripped off. The distributed nature of micro-grids is naturally adoptable by the inherent distributed profile of blockchain. Blockchain is operated using a distributed ledger, which is available to all participants to a common blockchain-based market scheme. The ledger is the receptacle of truths; each and every transaction is being verified, timestamped, ran through an advanced crypto machine, and added to the ledger. Another intrinsic secure feature of the ledger-based transactions is the fact that each transaction is traceable and is validated only by the transaction which preceded it. This architecture is basically sound for any security beach, since a hacker who gains access to the platform and wants to change a section of a ledger must change all the sections contained in the distributed ledger, to all peers. The success of this endeavor is virtually impossible.

The smart grids are basically arrays of technologically backed generation units, intelligently linked to the consumption units, which are the beneficiaries of wisely developed utilization strategies. These strategies sprung from the need of optimization of the usage of energy assets, in order either to maximize the revenues or minimize the losses, either energy or monetary wise.

The two apparently disparate paradigms show peculiar similarities with respect to the common management policies. By concealing the operation of the system (either physical or software) to a mathematical algorithm, designed to act as a business enabler for the lucrative peers, as well as a solid-rock guardian for possible intruders, the marriage between physical micro-grid (or smart grid) and the automated and collaborative system peer-to-peer blockchain network could be the best follow-up yet.

However, every technological breakthrough has both a bright side and a not-so-bright side. The latter one resides in the inherent energy-intensive nature of blockchain-based technology, which is an energy-intensive technology, which pays tribute to the constant care for locking down the information and continuous search for certitude in keeping the peers out of reach of hackers, activity which implies constant checking (every 10 seconds) of proof of work in every ledger at every peer, irresponsive to the continuous energy trading process. This feature of blockchain will need further improvement, since the energy quantities transitioned over micro-grid energy markets are small, and the overall gain of the community risks to become negative.

## 6. Conclusions

This chapter has discussed the framework of current electrical power system development into more elaborate and complex forms, which entails the *sine qua non* presence of the Internet. Energy efficiency domain is converging with the need for power quality improvement in the advent of smart grid solid-state technologies. This trend is supported by the seamless integration of IoT at both generation and consumption extremities, as well as their evolution toward micro-grids and smart grid, economically enabled by the blockchain-based micro-market schemes.

## Author details

Radu Porumb* and George Serițan
University POLITEHNICA of Bucharest, Faculty of Power Systems, Bucharest, Romania

*Address all correspondence to: radu.porumb@upb.ro

## References


[1] Energy management systems–Measuring energy performance using energy baselines (EnB) and energy performance indicators (EnPI)–General principles and guidance. ISO 50006:2014

[2] Energy management systems–Guidance for the implementation, maintenance and improvement of an energy management system. ISOCD3 50004. Standard under development, by Technical Committee ISO/TC 301 - Energy management and energy savings, due for publication in 2019

[3] Energy management systems–Energy management systems–Measurement and Verification of Organizational Energy Performance–General Principles and Guidance. 2014. ISO/DIS 50015

[4] Energy management systems–Requirements for bodies providing audit and certification of energy management systems; ISO/DIS 50003/2014

[5] Bertoldi P. Energy Efficient Equipment within SAVE; Activities, Strategies, Success and Barriers. Brussels: European Commission, Directorate General for Energy. Copenhagen. 2001

[6] Turner WC, Doty S. Energy Management Handbook. 6th ed. Boca Raton, USA: CRC Press; 2006. ISBN 9781466578289 - CAT# N10734

[7] Wollenberg BF. Power System Operation and Control. Boca Raton, USA: Taylor & Francis Group LLC; 2006. ISBN 9781439883204

[8] Chung YH, Kim HJ, Choe JW. Unified Power Quality Controller for the Microgrid System. Paris, paper C6-301: CIGRE; 2010

[9] Olivares DE, Canizares CA, Kazerani M. A centralized optimal energy management system for microgrids. In: IEEE PESGM, Minneapolis, Minnesota, paper 118. 2010

[10] Treatment of Losses by Network Operators ERGEG Position Paper for public consultation. European Regulators Group for Electricity and Gas. 15 July 2008. Ref: E08-ENM-04-03

[11] ABB Energy Efficiency Handbook. Power Generation Energy Efficient Design of Auxiliary Systems in Fossil-Fuel Power Plants. ABB 2009

[12] Arrillaga J, Watson NR. Power System Harmonics. Chichester, UK: John Wiley & Sons; 2003

[13] Energy performance of large power transformers ($U_m > 36$ kV or $S_r \geq 40$ MVA). Standard EN 50629:2014

[14] Panagiotis K, Lambros E, editors. Electricity Distribution, Intelligent Solutions for Electricity Transmission and Distribution Networks. Berlin Heidelberg: Springer-Verlag; 2016. pp. 27-61. DOI: 10.1007/978-3-662-49434-9

[15] Desmet J et al. Analysis of the neutral conductor current in a three phase supplied network with nonlinear single phase load. In: IEEE International Electric Machines and Drives Conference IEMDC 2001, Cambridge, USA. 2001

[16] IEEE Recommended Practices and Requirements for Harmonic Control in Electrical Power Systems. IEEE Std. 519-1992

[17] Chicco G, Postolache P, Toader C. Analysis of three-phase systems with neutral under distorted and unbalanced conditions in the symmetrical component-based framework. IEEE Transactions on Power Delivery. Jan. 2007;**22**(1):674-683

[18] Thomas H et al. Efficiency of superconducting transmission lines: An analysis with respect to the load factor and capacity rating. Electric Power Systems Research. 2016;**141**:381-391. DOI: 10.1016/j.epsr.2016.07.007

[19] Stemmple M, editor. Simplified undergrounding of 400 kV overhead lines with superconducting cable systems. CIGRE 2016. paper B1-308

[20] Status of Development and Field Test Experience with High-Temperature Superconducting Power Equipment. Technical Brochure. Working Group D1.15. CIGRE. paper 418; 2010

[21] Eremia M. Electric Power Systems. Electric Networks, E.A.R.; Bucharest, Romania: Romanian Academy Editing Press. 2007

[22] Eremia M, Liu C-C, Edris A-A. Advanced Solutions in Power Systems. IEEE Press, Wiley; 2016

[23] Siddiqui AS et al. Congestion management in high voltage transmission line using thyrister controlled series capacitors (TCSC). Journal of Electrical and Electronics Engineering Research. October 2011;**3**(8):151-161

[24] Akhil AA. DOE/EPRI 2013 Electricity Storage Handbook in Collaboration with NRECA. Albuquerque, New Mexico: Sandia National Laboratories; 2013

[25] Badr M, Ali Ahmed S, Emara KH. Reduction of energy losses in electrical distribution systems. Stockholm, Sweden: Proceedings of the 22nd CIRED. 2013. paper 0176

[26] Elsaiah S, Mitra J. A method for minimum loss reconfiguration of radial distribution systems. PESGM 2015, Denver. paper 002510

[27] IEEE Standard Definitions for the Measurement of Electric Power Quantities Under Sinusoidal, Nonsinusoidal, Balanced, or Unbalanced Conditions. IEEE Std. 1459-2010

[28] Helmi BA, D'Souza M, Bolz BA. Power factor correction capacitors. IEEE Industry Applications. 2015;**2**(5):78-84

[29] Nakamoto S, Bitcoin. A Peer-to-Peer Electronic Cash System. Available from: bitcoin.org/bitcoin.pdf [Accessed: 2018-12-09]

[30] Tapscott D, Tapscott A. Blockchain Revolution. Portfolio/Penguin; 2018. ISBN: 978-1-101-98014-9

## Chapter 6

# Sustainable Energy Model for the Production of Biomass Briquettes Based on Rice Husks in Peruvian Low-Income Agricultural Areas

*Juan Arévalo, Grimaldo Quispe and Carlos Raymundo*


## Abstract

An energy model focuses on the sustainability of environmental proposals that use clean biomass technology. In this case, briquette production seeks to generate socio-environmental development in agricultural areas contaminated by the burning of rice husks. However, this agricultural waste product has a large heating capacity and can be used as a raw material for briquette production, replacing conventional contaminant fuels such as firewood and reducing Peru's annual energy consumption by approximately 833,000 kg of $CO_2$ per year, considering the minimization of emissions from the felling of trees and the burning of rice husks. These rice husks are burned and generate pollutant gases, causing respiratory and pulmonary problems. Despite these negative effects, it is an agricultural waste product with great untapped energy potential and constitutes an opportunity to promote socio-environmental development based on economic valorization. The level of deforestation would decrease by approximately 2070 trees per year, 23% of a market population which consumes 10 kg of firewood per day. Unlike similar projects, briquette production sustainability may be achieved when economic, environmental and social aspects are included in energy model development, based on the application of clean technology and efficient management of energy supplies, such as husk supplies and corresponding briquettes.


**Keywords:** biomass, briquettes, circular economy, energy model, socio-environmental, heating capacity

## 1. Introduction

One problem in the rice industry is the accumulation of rice husks in high volumes, which are charred and thrown into rivers due to little interest in recycling these by-products for industrial sub-processes, thus constituting an opportunity to economically value an agricultural waste product within the value chain of paddy rice. The rice industry has a significant participation in the Peruvian economic sector, as the crop with the greatest contribution to agricultural development and GDP, producing approximately 44.7 million day wages and generating 161,300 jobs per year, representing its strong social and economic influence in rural areas. [1]. **Figure 1** shows the importance of the rice industry in Peru, which figures among the 20 countries with the highest production of paddy rice worldwide, processing a

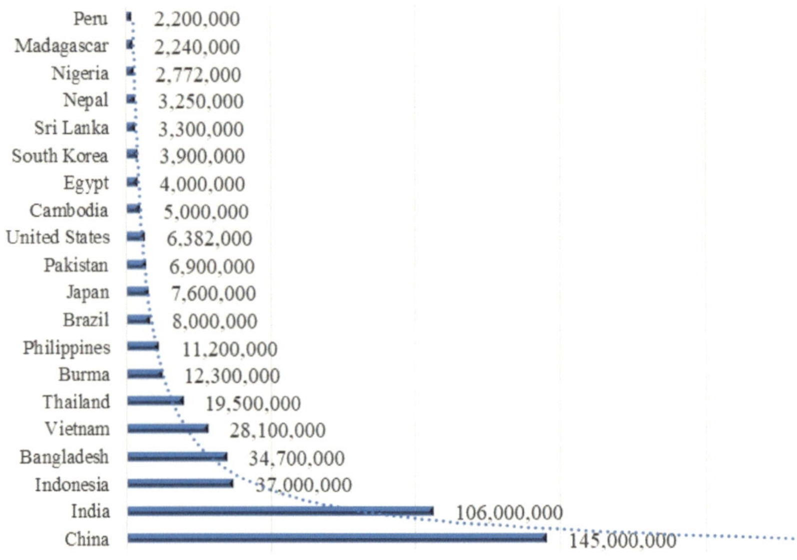


**Figure 1.**
*Hectares of rice husk planted during the 2013–2014 agricultural campaign in Peru.*

total of 2200 tonnes during the 2013–2014 agricultural campaign, the second Latin American country in this agricultural sector, since most of the production of this crop is found in Asian countries [2].

At the national level, San Martín is the region with the largest paddy rice area, annually reaching 86,053 hectares in the months of July and August [3]. Within this region, a rice industry company produces 45 tonnes of paddy rice daily; therefore, there is a large supply of rice husks around the mill or in the company's growing areas, many of which are dumped into rivers or nearby roads, harming the local environment and people.

The rice husk represents 20% of total paddy rice production [4]. Approximately 9 tonnes of this agricultural waste accumulates daily [5]. Thus, rice industry companies seek opportunities to recycle these waste products, with the purpose of converting their lineal production into a circular economy, based on the economic valorization of the rice husk through the elaboration of ecological products using clean technology, which does not generate additional costs with respect to obtaining raw material.

Within the concept of a sustainable energy model, several alternatives for rice husk recycling should be analyzed, which is why it analyzed the current situation of the area, where this agricultural waste product is abundant. One of the factors with the greatest environmental impacts is the use of firewood for cooking, exacerbating local pollution, which is currently damaged by burning rice husks. Therefore, biofuels are being developed through scientific research to replace firewood. It was determined that briquettes and pellet biofuels, also known as "ecological coal," have the highest energy efficiency. Both resources have been developed worldwide for projects with similar criticality to the current situation in San Martín regarding deforestation and high firewood consumption in rural and agricultural areas.

The present case determined that the briquette is the best option to replace firewood. Unlike pellets, briquettes may be used in domestic activities, notably in individual boilers, traditional ovens, or fireplaces. These briquettes generate greater opportunities for new markets, since they can be elaborated in different forms and sizes, unlike pellets which are necessarily cylindrical and smaller [6].

Briquette development is by no means a new technology in agricultural areas, where there are many types of biomass or resources for biofuel material. Projects with similar characteristics have already been implemented, from an environmental perspective and with a vision to maximize energy resource use. However, some did not achieve successful results, and not necessarily because of product quality but because of the way in which the proposal was developed. Others were successful but could have had greater economic impacts.

Among the most important aspects regarding briquette proposals is the determination of raw material costs assumed by rice companies and the way in which it would impact their current budget. For example, "Corinay Briquettes" is a producer, exporter, and marketer of coal briquettes, whose initial objective was to exploit the abundance of this resource in rural areas. In addition, it would substitute domestic use of firewood, thus reducing the rate of deforestation in the region [7]. One of the main problems is the increase in the cost of coal, since "Corinay Briquettes" increased its installed capacity to 5000 tonnes per year [8], due to the large supply of briquettes that exists in rural areas, where production is concentrated. In comparison with the present proposal for briquette production under a sustainable energy model, the difference is that the rice husk will have zero costs, at least in the first years of operation.

Another important aspect from other projects is the determination of different briquette production programs. For example, the company "North Wood" established different production scenarios for sawdust briquettes, focusing its business on an analysis of supply and demand, providing knowledge regarding the number of briquettes necessary to satisfy current consumption of the population using firewood in a certain area [9]. Thus, energy model sustainability will focus on the fulfillment of short- and medium-term demands so that different production scenarios may be established which respond to demand variability for this innovative ecological product, as it may succeed based on market acceptance.

Energy model sustainability also focuses on societal acceptance and perception of briquettes. For example, the company "Eco Amazonia" did not succeed in its coconut husk briquette business because the biofuel was sold under the name of "Ecocarbón," with a coal shape and color. For this reason, society perceived that it was not an ecological product, since it was composed of charred coconut husk, which would later generate a large amount of volatile matter [10]. Therefore, the present proposal considers the inclusion of non-charred rice husks for briquette production.

Additionally, one of the main factors within the energy model is determining a specific place where the proposal will be developed. As mentioned above, the San Martín region is an ideal place for briquette production due to its great potential for growing rice husks coupled with high levels of firewood consumption. In this region, the sector with the highest quantity of processed rice is central Huallaga, with a total of 73,343 tonnes, divided between the provinces of Picota, Huallaga, Bellavista, and Mariscal Cáceres. However, the high volume of rice husks is not the only factor for determining the location; it is also necessary to consider environmental awareness of disadvantaged residents. For example, the municipality of San Hilarión, located in the central sector Huallaga, established Municipal Ordinance No. 013-2004/MDSH/A, which prohibits the burning of husks by mill owners. The economic sanction is valued at 2 UIT [11]. The opportunity to invest in clean technology becomes an obligation for several mills in the area; hence, they seek alternatives to recycle rice husks, economically valuing the produced biomass. **Figure 2** presents the regions with the highest rates of environmental pollution due to rice husk burning.

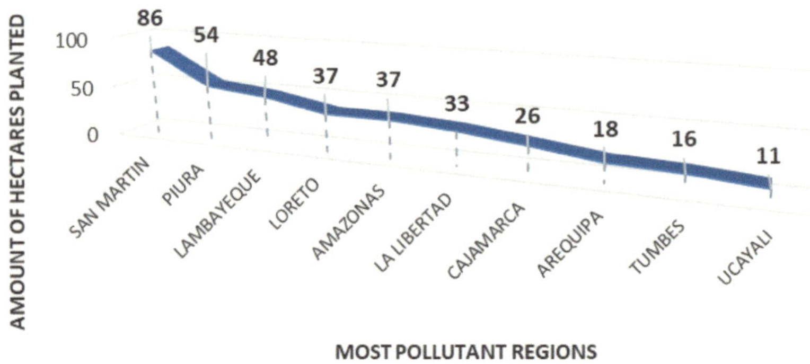


**Figure 2.**
*Environmental pollution due to rice husk burning during 2013–2014 agricultural campaign in Peru.*

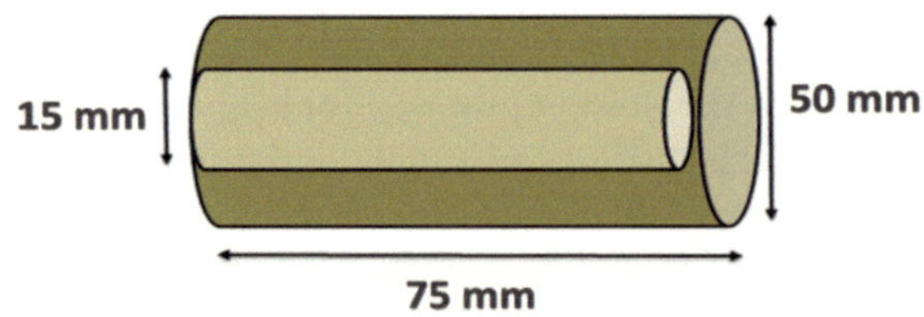


**Figure 3.**
*Dimensions of rice husk briquettes.*

As mentioned above, the briquette is not a new technology, so determining briquette design will depend on product use and specifications of the briquette machine. For industrial processes or businesses, lengths vary between 300 and 1000 mm, for producers length varies between 100 and 500 mm, and for domestic sector, length varies between 30 and 80 mm. **Figure 3** shows the ideal prototype of briquettes for domestic use.

The inclusion of an interior hole will endow the briquette with greater oxygenation capacity but could increase volatile matter, so the shape of the product will depend directly on its use, be it industrial or domestic. Another variable is the market approach to briquette production, since it will not necessarily be used as fuel but can also act as a heating resource in locations with low temperatures or may be exported to the European market. Density is another main characteristic, since as it becomes denser; less volume will be occupied, which will mean easier handling, optimum storage, and easier transport, compared to firewood. Its weight should be 1000 kg/m3 [12], and this depends mainly on rice husk density and the pressure exerted by briquette machines. Finally, humidity directly influences heating capacity, for they contain a large percentage of moisture and the energy released is lower during combustion, causing evaporation to consume heat. Humidity should vary between 8 and 10% [13].

## 2. Method

Within the methodology, the goal is to develop the energy model, which is mainly based on the use of renewable energy and the search for energy efficiency, with the purpose of ensuring sustainability in the proposal, satisfying economic, environmental, and social aspects.

The model begins with the search for alternatives for recycling biomass as an alternative resource to firewood, which is why briquettes emerge as the ideal biofuel in the environmental proposal development. Briquettes are formed from the conglomeration of various types of waste, may it be forestry, agricultural, or industrial [14], mixed with binders such as cassava starch or bentonite, in order to optimally compact the mixtures. The choice of agricultural waste or biomass as raw material for the briquette was rice husk, due to the high rate of charred husk on location, which will allow for a large supply of this waste, reducing the population's social cost.

Having selected the clean technology, it is necessary to perform a technical evaluation of the briquette based on energy efficiency, which includes calculations of heating capacity and efficiency during combustion. The rice husk briquette has 4040 kcal/kg heating capacity, that is, the amount of heating capacity used per kilogram burnt [15]. Its combustion energy efficiency obtained a value of 80.39%; in comparison to firewood, fewer kilograms of briquettes are needed to heat or prepare food in a shorter time. Additionally, it is important to set out the development method within the technical product evaluation, in order to meet certain technical requirements, such as moisture measurement, through the drying method, which uses the "Colombian technical standard for domestic use briquettes," permitting a value between 9 and 10%. Furthermore, there are two important aspects that were investigated to obtain an optimal briquette, based on adequate biofuel combustion: granulometric analysis and agglomeration analysis. Granulometric analysis is used to achieve a resistant briquette with sound composition, analyzing different sizes of husk particles with metal sieves. On the other hand, agglomeration analysis, using cassava starch as a binder, will be important in balancing mixture resistance and moisture. Various forms have been used, the first of which compresses the briquette with a piston and the second by manual press compression. Thus, a briquette with high quality standards, based on the percentage of moisture, ash, and volatile matter, was obtained.

To ensure the energy model—designed specifically for biomass briquette production proposals—is developed successfully, it is important to establish an integrated process system, starting from the supply and demand analysis process, to briquette commercialization and the opening of new markets. In this analysis, a market study is conducted to determine the population that uses firewood for cooking, as well as the supply of rice husk quantities in the mills. In addition, the sales price is determined, including the profit percentage that the company will earn and estimations of the final proposal budget. Then, it will be necessary to determine the logistics of the proposal's supply chain, beginning with alternatives for rice husk and cassava starch supply, to later produce various production programs based on scenarios of low, medium, and high demand for briquette sales.

Once the production scenarios are determined, a processing plant must be designed, with enough physical capacity to store material for a high demand, considering probable openings to new markets or potential customers, including adequate distribution of machinery, materials, and other physical production resources. The supply chain culminates with the commercialization of briquettes in rural areas where the product will be consumed, considering the necessary resources and transport costs for correct development. For example, the company or mill should include the transport of cassava starch from agricultural areas to the processing plant or the most efficient form of transport for the commercialization of briquettes and rice husks.

Finally, the energy model seeks to be sustainable by ensuring process operability, so the economic, environmental, and social impacts of the briquette production proposal are analyzed. Regarding the economic impact, companies plan to economically value agricultural waste products, which will subsequently generate additional income, entering new briquette markets in both domestic and industrial sectors. Regarding the social impact, quality of life was improved, since inhabitants' respiratory and lung disease incidence rates decreased, from the reduction of $CO_2$ produced by burning rice husks and firewood in the domestic sector. **Figure 4** shows the energy model proposed for rice husk briquette production.

Finally, environmental impact was measured by the reduction of $CO_2$ produced by burning rice husk in paddy fields or around the city and by the minimization of greenhouse gases (GHGs) from substituting wood for briquettes in food preparation.

## 2.1 Operationalization

As part of process operationalization, a rice husk supply analysis is necessary, which ensures the raw materials meet production demands. For example, San Martín is the region with the highest economic participation in terms of rice industries in Peru, representing 18.49% of annual production in the 2013 agricultural campaign, with a total of 563.99 tonnes of paddy rice. Of the total amount produced, 20% will be converted into husk, meaning large volumes of this agricultural waste product will be burnt, due to limited recycling. **Figure 5** shows the various current uses of rice husk [16], reflecting its potential demand for companies or mills within the regional market. In many agricultural areas of the San Martín region, policies have been established to raise awareness and encourage recycling in other activities, regulated by municipal ordinances prohibiting burning. In addition, from a survey of 100 families in the region, the study concluded that 70% of the population would be willing to substitute firewood for an ecological product, 50% would pay a price equal to that of firewood, and 70% agree that briquettes should be delivered to their homes.

On the other hand, energy model sustainability revolves around the acceptance of demand variability in any successful scenario for briquette sales. For this reason, certain production programs were designed based on briquette machine production capacity. For example, if the market focus is 23% of the firewood-consuming population, mills or companies should produce 69,000 briquettes per month for a total of 92 families, using 13,248 kg of rice husk and 552 kg of cassava starch. Within a supply–demand analysis, the final stock estimate is 36,000 briquettes, which may be distributed monthly to low-income households or be sent to poulterers or bakeries as an alternative source for embers in their ovens.

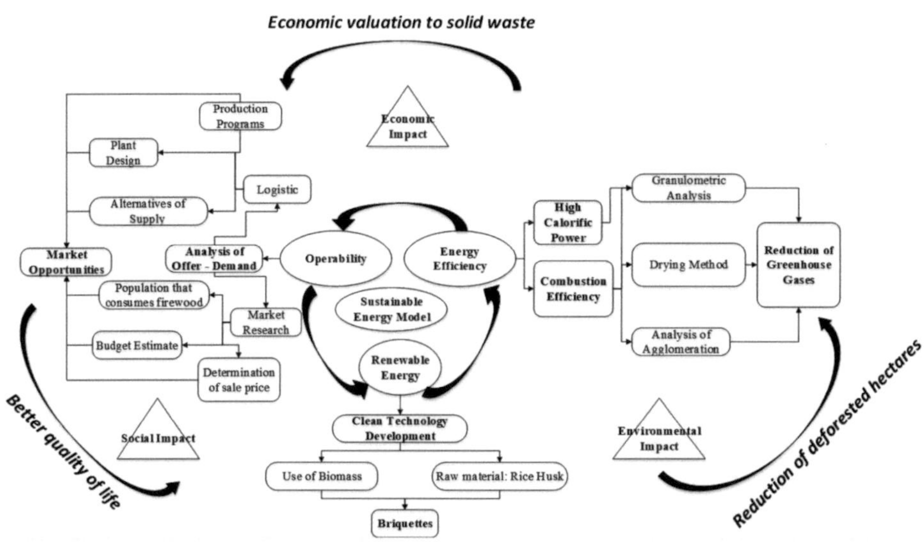


**Figure 4.**
*Sustainable energy model design.*

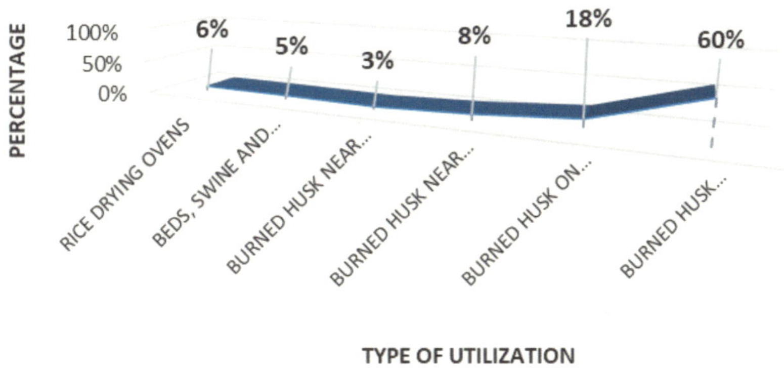


**Figure 5.**
*Utilization of rice husks in agricultural areas of the San Martín region.*

However, if at the end of the year a briquette production program is not desired, a second production program can be designed, without the need to change the market focus completely. For example, 20% of the market should initially be addressed between January and July, with a monthly production of 60,000 rice husk briquettes. Subsequently this would increase by 40% from July to December, to supply a total of 112 families with 84,000 briquettes, using the briquette machine's maximum production capacity. Including the final inventories from January to June, this would deplete resources to zero in December. It bears noting that the second production scenario must consider approval and sales success during the first semester.

Note that the proposed briquette machine is of Italian origin—model E60, ECO by Prodeco—which has been used in similar briquette production projects. Its processing capacity is 60 kg of cassava starch and rice husk mix, for a total production of 300 briquettes per hour. Thus, the briquette machine processes a maximum of 2400 briquettes per day. In addition, these machines work 8 hours a day, transforming 1 kg into five briquettes for a market which consumes an average of 5 kg per family. Therefore, a total 25 briquettes per family must be produced for daily consumption, considering the yield equivalence between husk briquettes and firewood.

The briquette production program must consider a processing plant design (see **Figure 6**), with the purpose of strategically distributing the machines and resources corresponding to briquette production. For example, mixing and grinding areas should be separated, because smoke from the grinder must not come into contact with the mixing process of binder and husk.

The production process begins with the selection of raw material, which is comprised of two stages. First, sand, spikes, and residue of different sizes are eliminated using sieves, in the granulometric analysis. The "weighing" process follows, wherein the various resources used during the entire production process are weighed. Next, the "mixing" process, by means of the agglomeration method, combines the rice husk in different quantities. Finally, the mixture is compacted by the briquette machine and passed through the "drying" area, to then be packaged for sale.

## 2.2 Method

One of the most relevant methods for briquetting is the granulometric method, which is a process of compaction or briquetting by using a piston and a press to reduce mixture structures. However, a briquette machine may also be used,

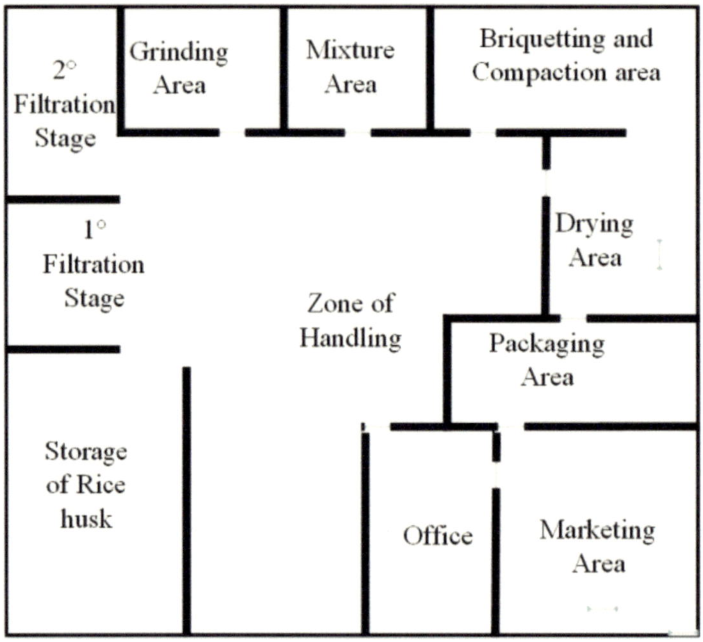


**Figure 6.**
*Design of briquette processing plant.*

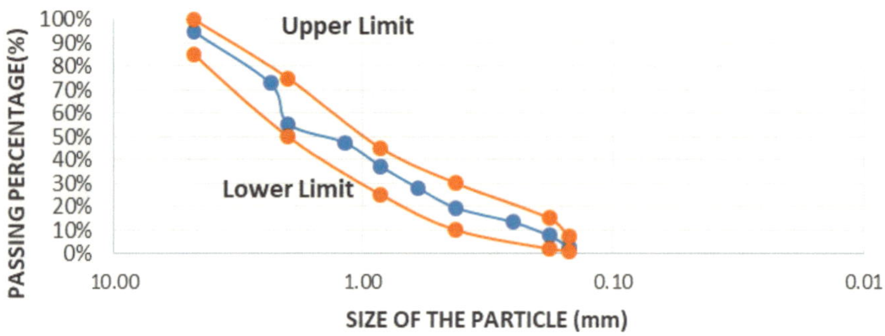


**Figure 7.**
*Granulometry of rice husk particle composition.*

optimizing man-hours and increasing production rates. Development methods were performed in the laboratory at the Peruvian University of Applied Sciences (UPC).

The granulometric method is important for proper mixture compaction, since there is a range indicating optimum mixture state; for example, the lower limit indicates a mixture with a larger proportion of finer rice husk, whereas the upper limit indicates a mixture with larger grains. However, if the mixture falls outside the established range, the fine grain mixture will hinder oxygen flow and combustion, whereas if it exceeds the upper limit, the excess oxygen will produce more pollutant gases [17]. **Figure 7** shows the ratio of rice husk particle size necessary for adequate composition. Generally, particle size ranges from 0.10 to 3.00 mm. Composition depends directly on use, for the mixture can be composed of small or large particles, but must fall within this range [13].

The granulometric distribution of the best briquette prototype elaborated in the UPC laboratory is shown in **Figure 8**. This prototype had the greatest energy efficiency, both in terms of correct combustion and heating capacity. This briquette

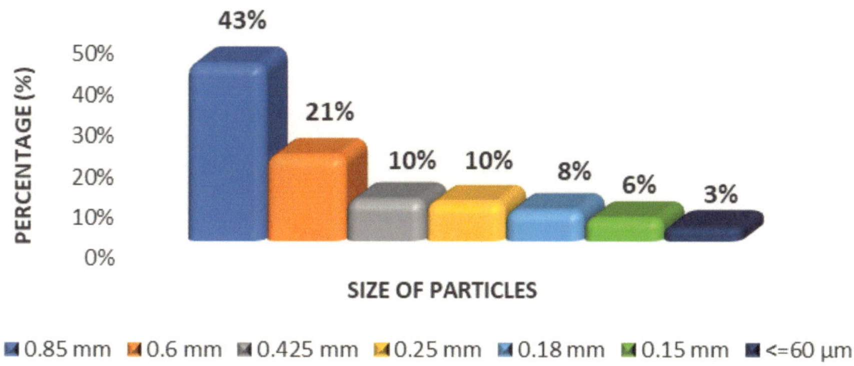


**Figure 8.**
*Granulometric distribution of rice husk particles.*

is composed of 0.850 mm particles, as well as smaller particles, making it perfect for briquette compaction using an experimental artisanal method. However, as mentioned above, briquette composition should include particles of different sizes, especially for industrial production, because the briquette machine can compact larger rice husk grains correctly, with an average size of 2.36 mm [18].

Another method within the energy model that will provide sustainability from a technical perspective is agglomeration, which will generate value to briquette production. Agglomeration is the initial mixing stage of cassava starch and water. One liter of water and 200 g of cassava starch were used for this experiment, producing approximately 15 briquettes. The agglomeration process for this type of briquette production was the following, considering the percentage of each input in each stage:

1. Twenty-five percent of cold water is mixed with 200 g of cassava starch and stirred for 2 minutes.

2. Seventy-five percent of the total water is boiled to 100°C.

3. While boiling, the mixture is stirred until adhesive.

4. The adhesive mixture is then added to the rice husk container.

The agglomeration process can be seen in the left section of **Figure 9**, where the entire sequence of general rice husk briquette production processes is also shown, beginning with the "selection" of raw material, comprised of two stages, wherein sand, spikes, and residue of different sizes are eliminated by filtering. The "weighing" sub-process follows, in which the various resources used throughout the production process are weighed. Next, the "mixing" sub-process is performed, by agglomeration of water and cassava starch, which is later combined with rice husk in different quantities. The mixture is then compacted, either by a briquette machine for industrial purposes, or a hydraulic press for experimental artisanal purposes. Finally, the briquettes pass through the "drying" area and are packaged for sale.

It is important to note that both methods—granulometry and agglomeration—produce optimum values regarding combustion efficiency and heating capacity, thus both variables will be measured through the "boiling water test" (BWT), which measures the amount of energy transferred from the biofuel to a pot or container with a certain volume of water [19]. Below are some basic characteristics of the BWT:

- Sufficient quantities of water and fuel for the experiment. The fuel must be uniform and completely dry.

- The volume of cold water must be at least 10% of the total water used for the experiment.
- Temperature is measured at boiling point, because at that moment, temperature can no longer vary nor the amount of transferrable energy.

Another BWT characteristic was biofuel moisture content calculation. This observation is related to that mentioned by Michael Lubwama and Vianney Yiga, in the paper "Characteristics of Briquettes Developed from Rice and Coffee Husks for Domestic Cooking Applications in Uganda" in which they state that the moisture content of any type of biomass must oscillate between 10 and 15%, with a compression force of 230 MPa. In addition, humidity values should oscillate between 9 and 10%, to obtain efficient combustion with fewer gas emissions. However, biomass compression is not the only moisture reduction process; a drying process must also be carried out, in which an electric heater is used to reduce humidity before the mixing process. Calculations referring to different means of calculating moisture content will be shown in the following formulae [20]:

$$\text{MOISTURE \% (wet)} = \frac{\text{(Fuel Mass)wet} - \text{(Fuel Mass)dry}}{\text{(Fuel Mass)wet}} \text{ x } 100 \tag{1}$$

$$\text{MOISTURE \% (dry)} = \frac{\text{(Fuel Mass)wet} - \text{(Fuel Mass)dry}}{\text{(Fuel Mass)dry}} \text{ x } 100 \tag{2}$$

$$\text{MOISTURE \% (wet)} = \frac{\text{MOISTURE (dry)}}{\text{MOISTURE (dry)}+1} \tag{3}$$

Eqs. 1 and 2 are important in analyzing briquette and firewood moisture content in different states, but the equation that was used, which is more frequent in this type of experimental method, corresponds to a dry base humidity calculation, due to better energy efficiency during combustion. Eq. (3) shows the different moisture content calculations.

As mentioned previously, rice husk moisture content is determined using an electric stove, after having passed the grinding process. Subsequently it is weighed and compared with the value obtained in drying. In Eq. 4 it observes the relationship between rice husks mass in different states of humidity:

$$\text{MOISTURE \% (rice husk)} = \frac{\text{Initial Mass}}{\text{Initial Mass} - \text{Final Mass}} \text{ x } 100 \tag{4}$$

As mentioned above, humidity calculations are important for optimal briquette combustion, based on combustion efficiency and heating capacity. In the study "Briquette Production for Use as a Power Source for Combustion, Using Charcoal Thin Waste and Sanitary Sewage Sludge," the authors describe the relationship between heat lost from the fuel and the heat of the water; thus Eqs. 5 and 6 show the calculations for both variables, including some theoretical values [21].

A 150 convective factor ("H" air-water) was considered for heat loss calculations, denominated "Q lost," since it is the most commonly used value with respect to thermal experiments [22]. Regarding water heat, denominated "Q water," the theoretical value for the specific heat of the water, was considered to be 4.18 J/g°C under normal conditions [23]; this value is represented by the "Cp." In addition, the container in which the test was performed had a diameter of 0.12 m and a height of 0.1 m, so the volume container is denominated "A." The equations necessary to estimate briquette heating values are shown below:

$$\text{Q lost (J)} = \text{A} \times \text{H} \times (\text{final temperature} - \text{initial temperature}) \tag{5}$$

$$\text{Q water (J)} = \text{Cp water} \times (\text{final temperature} - \text{initial temperature}) \times \text{water mass} \quad (6)$$

**Table 1** presents values used in the experiment, considering initial temperatures under normal conditions and boiling point as final temperatures.

The difference between these equations allows us to calculate the approximate amount of heat emitted during combustion, which is divided by the charred mass of the briquette. Eq. 7 calculates heating capacity (HC):

$$\text{HC (kcal/kg)} = (\text{Q water} - \text{Q lost})/\text{burned mass} \quad (7)$$

The equation developed by Estela Assureira [24] was used to calculate heating capacity, in which the resultant briquette mass is related to the mass of the inputs used for its production and the ash content representing each component. In Eqs. 8 and 9, the mass of the inputs added to the composition of the briquette is called "M bc." Note that in the experiment, the addition of another binder is necessary. The proportion of this input and its percentage of ash must be included in Eq. 9.

Finally, it bears mentioning that from the equations shown above, it was possible to calculate estimated heating capacity, humidity, and efficiency values, without the need for heating pumps or other laboratory instruments. However, if the focus

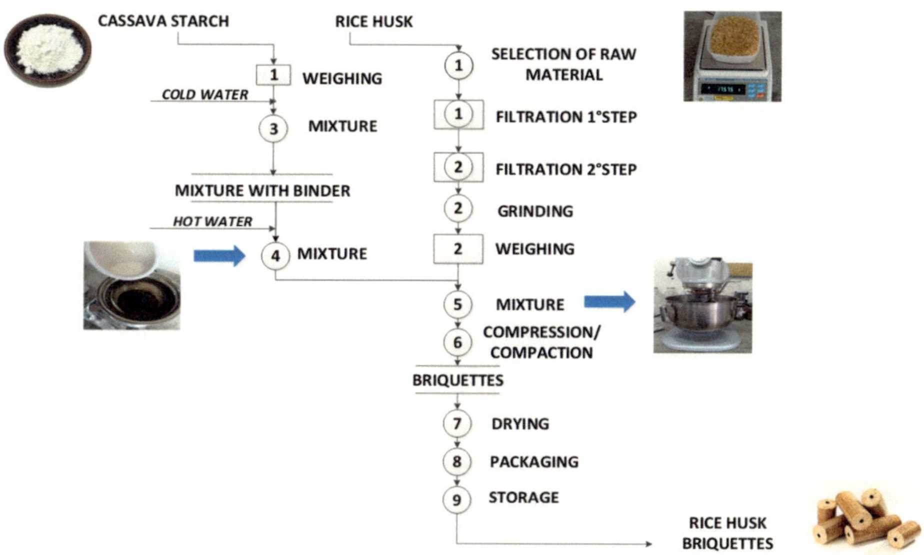


**Figure 9.**
*Rice husk briquette production process.*

| Characteristic | Value |
|---|---|
| Final temperature (°C) | 100 |
| Initial temperature (°C) | 22 |
| Cp water (J/g.°C) | 4.18 |
| Container volume ($m^3$) | 0.012 |
| H air–water ($W/m^2.K$) | (20–300) |

**Table 1.**
*Experiment characteristics.*

of the briquette production proposal was industrial and with a business vision, it would be necessary to use a briquette machine and sophisticated measuring instruments.

$$\text{combustion efficiency (\%)} = \text{(initial mass - final mass)/mass bc} \quad (8)$$

$$\text{mass bc (kg)} = [\text{mass husk} \times (1 - \%\text{ash})] + [\text{mass binder} \times (1 - \%\text{ash})] \quad (9)$$

## 3. Results

Experiment results will be analyzed in two ways. The first will be based on the energy analysis of the briquette prototypes and the comparison of their physical–chemical properties with firewood through agglomeration and granulometry methods. The second analysis corresponds to the environmental impact in low-income agricultural areas in Peru, where firewood is used, from the production of rice husk briquettes.

### 3.1 Energy analysis

The experiment was based on the preparation of various briquette prototypes, considering size, shape, and composition as relevant characteristics for each type, to obtain the greatest energy efficiency—that is to say, a similarity in heating capacity and combustion efficiency.

**Table 2** shows that the best briquette prototype is 1, which is referred to as "BR 1," with a heating capacity of 4040 kcal/kg and a combustion efficiency of 80.39%. This briquette is composed of 80% rice husk and 20% cassava starch. Other ingredients are not suitable for making briquettes, such as rubber which increases humidity or bentonite which results in low compaction levels. **Table 2** also shows the physical-chemical characteristics of each briquette prototype, represented by the nomenclature "BR."

Within the results presented in **Table 2**, density is the physical-chemical characteristic relevant to briquette production methods, due to the type of machine used for compaction. For this experimental process, in which a hydraulic press was used, density was greater than that of a briquette machine, due to the greater compression force that machines have (an average theoretical value of 350 kg/m$^3$, based on statistical data regarding the impact between compression capacity and this type of material) [25]. The results obtained in Ref. [26] showed a very low value of the density. Adding a binder increases moisture content, reducing combustion efficiency, since it is proportional to the increase in density. In addition, they mention an important aspect regarding rice husk transport costs, stating that this cost will be higher due to reduced availability [26].

Agglutinants should be added to artisanal briquettes to better compact the husk and cassava starch mixture. In addition, the cost of transporting the briquettes is always nominal. For the present case, briquette density was 678.7 kg/m$^3$.

As well as comparing heating capacity between different briquette prototypes, it will be necessary to compare the physical-chemical properties with those of firewood, to determine the advantages of using briquettes. In **Table 3**, a greater heating capacity is observed compared to firewood, since firewood loses a large part of heat energy due to certain properties. Briquettes, however, burn their initial

| Characteristic | Components | BR 1 | BR 2 | BR 3 | BR 4 | BR 5 | BR6 |
|---|---|---|---|---|---|---|---|
| Composition | Rice husk | 80 | 90 | 75 | 90 | 80 | 80 |
| | Yucca starch | 20 | 0 | 15 | 10 | 10 | 0 |
| | Bentonite | 0 | 10 | 5 | 0 | 10 | 0 |
| | Rubber | 0 | 0 | 5 | 0 | 0 | 20 |
| Shape | | | | Cylindrical | | | |
| Diameter (mm) | | 73 | 73 | 53 | 53 | 53 | 73 |
| Height (mm) | | 22 | 28 | 35 | 30 | 30 | 37 |
| Humidity (%) | | 9.63 | 10.97 | 8.23 | 8.23 | 9.46 | 10.86 |
| Bulk density (kg/m$^3$) | | 678.7 | 654.7 | 604.3 | 409.5 | 906.5 | 963.7 |
| Time of ignition (min) | | 8 | 12 | 10 | 11 | 12 | 10 |
| Heating power (kcal/kg) | | 4040 | 4010 | 3745 | 4020 | 4010 | 3500 |
| Combustion efficiency (%) | | 80.39 | 78.13 | 79.17 | 76.29 | 77.14 | 71.29 |

**Table 2.**
*Briquette prototype analysis.*

| Characteristic | Unity | Rice husk briquettes | Firewood |
|---|---|---|---|
| Heating power | kcal/kg | 4040 | 4010 |
| Bulk density | kg/m$^3$ | 860 | 820 |
| Ash | % | 1.50 | 0.92 |
| Humidity | % | 9.0 | 17.4 |
| Fixed carbon | % | 15.6 | 16.8 |
| Volatile matter | % | 86.5 | 82.2 |
| Combustion efficiency | % | 80.39 | 70.40 |

**Table 3.**
*Comparison of energy characteristics for both fuels.*

properties until completely consumed. The briquette has a combustion efficiency value of 80.39%, approximately 10% more than firewood, meaning less fuel would be needed for food preparation.

**Figure 10** shows the briquette prototype 1, denominated "BR 1," which obtained a heating capacity value of 4040 kcal/kg and a combustion efficiency of 80.39%.

### 3.2 Environmental impact analysis

Within the energy model, environmental impact will be measured based on the $CO_2$ reduction from the use of rice husks in briquette production, as well as gases emitted by firewood for food preparation and the amount of $CO_2$ produced by cutting down trees.

The first source of $CO_2$ emissions is the number of hectares deforested due to firewood production; thus, substituting firewood with briquettes will reduce greenhouse gas emissions. The amount of firewood obtained per tree cut was calculated

**Figure 10.**
*"BR 1" rice husk briquette prototype 1.*

in a study by the National University San Martín (UNSM) in Tarapoto [27]. Eq. 10 shows the volume of an average tree in the San Martín region, which is 0.32 $m^3$. Note that the height of the tree is denominated as "commercial H," the coefficient value for San Martín is denominated "Cf," and "AB" was calculated in Eq. 10:

$$\begin{aligned} \text{Volume}\left(m^3\right) &= D^2 \times 0.7854 \times \text{commercial H} \times \text{Cf} \\ \text{Volume}\left(m^3\right) &= 0.302 \times 0.7854 \times 7 \times 0.65 \\ \text{Volume}\left(m^3\right) &= 0.3216 \end{aligned} \tag{10}$$

On the other hand, based on the estimated volume of firewood obtained from a tree, which is 0.32 $m^3$, it was possible to calculate the estimated number of trees that would be deforested. Information was taken from a study by "Tienda Biomasa" of the Spanish company Leñas Oliver SL, which for the sale of its ecological products such as briquettes, pellets, and other biofuels calculated that 1000 kg of firewood is equivalent to 2$m^3$ [28]. Therefore, if a specific briquette production proposal focuses on 23% of the market, in a population that consumes 27,600 kg of firewood monthly, the volume of firewood obtained would be 55.20 $m^3$. From the value obtained in Eq. 10, it was estimated that 173 trees would be saved monthly.

In the analysis of trees saved from the use of briquettes, it is important to consider the number of forest hectares that will be protected. In this case, the trees of the San Martín region belong to 50-year secondary forests, according to data extracted by the Moyobamba Forestry Office [29]. Thus, 6 fewer hectares will be deforested per year, according to characteristics in **Table 4**.

**Figure 11** shows the exponential relationship between the number of trees saved and the amount of $CO_2$ emissions that each of them represents. From a study on the environmental impact of firewood, by Peruvian researchers Torres, H. and Polo, C., with the collaboration of scientists Seifert, D. and Neuoetting, D., it has been found that 1 kg of firewood emits 1.83 kg $CO_2$, since half the wood's mass is carbon (C) and its relation with the molecular weight of $CO_2$ is 44/12, thus 1 kg of firewood produces 0.5 (44/12 kg of $CO_2$) = 1.83 kg of $CO_2$ [30].

As mentioned above, the number of deforested hectares and $CO_2$ emissions will depend directly on the type of market on which briquette production will focus. In this case, the monthly consumption of 23% of the population of San Hilarión is 27,600 kg of firewood, which represents monthly $CO_2$ emissions of 50,508 kg. Therefore, 33,120 trees are estimated for a 5-year period; that is to say, the use of rice husk briquettes would reduce $CO_2$ emissions by almost 10,000 kg.

Finally, in addition to reducing $CO_2$ emissions by protecting trees, $CO_2$ emissions will further be reduced with respect to burning rice husks in cultivation areas or around the city, due to the fact that a considerable amount of this agricultural waste product will be recycled in briquette preparation. From a physical-chemical rice husk analysis by the research group *Gestión Ambiental Sostenible* (Sustainable Environmental Management), formed by environmental engineers Aberlardo Prada and Carol Cortés of the University of Llanos, 1 kg of carbonized rice husk is equivalent to 1.43 kg of $CO_2$ [31]. Thus, from the number of briquettes produced, the $CO_2$ reduction from recycling rice husks can be calculated. **Figure 12** shows the relationship between the amount of husk used to make briquettes and the reduction in $CO_2$ emissions.

| Type of forest | Diameter type | Trees/ha |
|---|---|---|
| Secondary forest (50 years) | 20–30 cm | 340 |
| | Over 30 cm | 100 |

**Table 4.**
*Characteristics of forest resources in San Martín.*

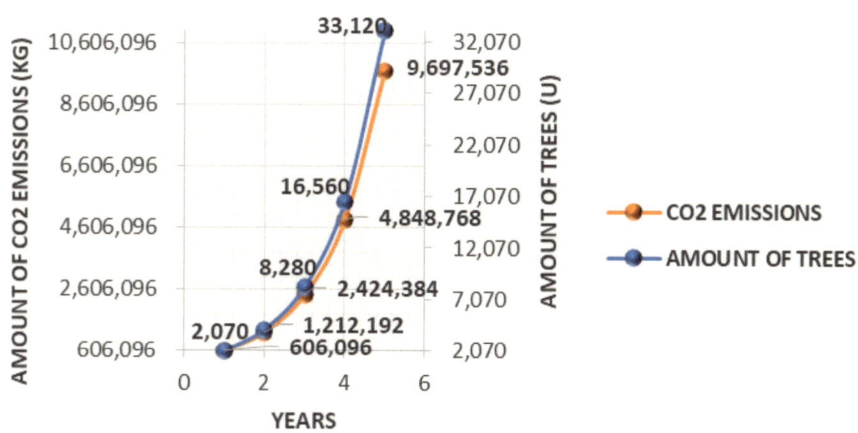


**Figure 11.**
*Amount of $CO_2$ emissions avoided by not cutting trees.*

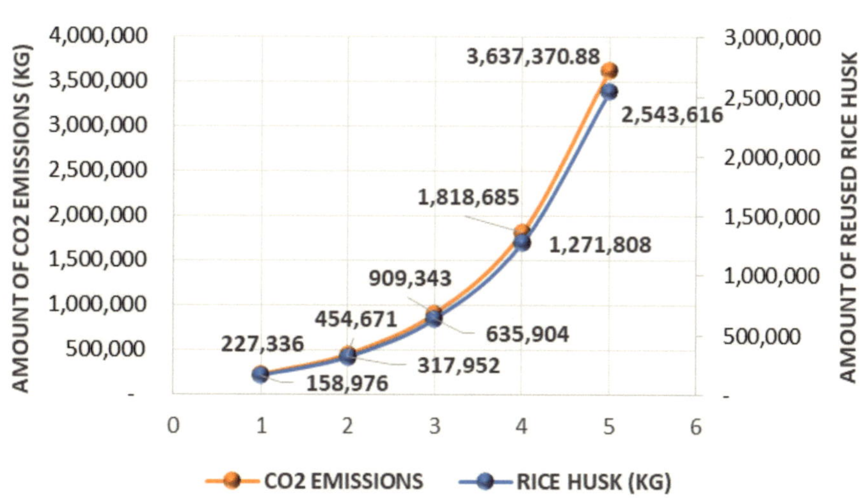


**Figure 12.**
*Amount of $CO_2$ emissions avoided by recycling rice husks.*

In this way, social costs on the population will be minimized, since there will be fewer respiratory and pulmonary diseases due to the reduction of $CO_2$ emissions from felling trees and burning rice husk. It was estimated that in 1 year, $CO_2$ emissions could be reduced by 833,000 kg.

## 4. Conclusions

The energy model design will provide sustainability to a specific proposal for the production of rice husk briquettes, from both an energetic and an environmental perspective, in which a quality product based on renewable energy is proposed. Additionally, from an economic and efficient perspective, the supply chain of all the resources corresponding to the production process is optimized and managed efficiently.

Furthermore, briquette production will generate the opportunity to recycle agricultural waste products from the rice industry, which is currently obliged to make use of them without burning in the vicinity of the city or agricultural areas, due to economic sanctions established by the municipal ordinances in the San Martín region. In addition, rice husk will be economically valued within the paddy rice value chain, as new markets will be opened, positioning briquettes as a viable alternative to firewood. Advantages such as heating capacity and reduced greenhouse gases are key, minimizing the rate of pulmonary and respiratory diseases in the population.

From an experimental analysis of different types of rice husk briquette, performed in the laboratory at the Peruvian University of Applied Sciences (UPC), an ideal briquette prototype was obtained with a combustion efficiency of 80.39%, reducing the amount of kg used for food preparation by 30%. As for heating capacity, the briquette obtained a value of 4040 kcal/kg, which is greater than firewood. The briquettes made the water reach boiling point before firewood, which registered a temperature of 70° C when the briquette had already reached 100° C.

Finally, rice husk recycling will generate a circular economy within the paddy rice value chain, promoting an improved environmental culture in society through the development of clean technology, which focuses on the reduction of greenhouse gases, amounting to approximately 833,000 kg of $CO_2$ per year, considering the protection of forest hectares and rice husk briquette production.

## Author details

Juan Arévalo[1], Grimaldo Quispe[1] and Carlos Raymundo[2*]

1 Faculty of Engineering, Universidad Peruana de Ciencias Aplicadas (UPC), Lima, Peru

2 Department of Research, Universidad Peruana de Ciencias Aplicadas (UPC), Lima, Peru

*Address all correspondence to: carlos.raymundo@upc.edu.pe

## References

[1] García H. Barreras para el Desarrollo de la Bioenergía. Matriz Energética en el Perú y Energías Renovables. Lima: Fundación Friedrich Ebert (FES) en colaboración con Derecho, Ambiente y Recursos Naturales (DAR); 2013

[2] Orrego Moya R. Estado del arte y novedades de la Bioenergía en el Perú. La Bioenergía en América Latina y el Caribe. Santiago: Organización de las Naciones Unidas para la Alimentación y Agricultura (FAO); 2013. pp. 341-375

[3] Programa Nacional de Conservación de Bosques para la Mitigación del Cambio Climático. En Análisis de la Realidad Propia. Lima: Editorial of the Environment Ministry; 2013. pp. 18-45

[4] Assureira E. Rice husk, an alternative fuel in Peru [Abstract]. Biomass and Coal Research Programme. Lima: E-Publishing of the Pontifical Catholic University of Peru; 2002

[5] Muro J. El Arroz: Principales aspectos de la cadena agroproductiva. Lima: E-Publishing of the Direction General of Agricultural Competitiveness; 2015

[6] Marcos MF. Pélets y Briquetas. 1994. pp. 54-62. Extracted on: http://www.informadera.net/uploads/articulos/archivo_2293_9990.pdf

[7] CORINAY. Briquetas Corinay. 2008. Extracted on: www.corinay.com/

[8] Garcia M. Diseño de Proceso y Planta Piloto para Fabricación de Briquetas de Aserrín. In: Apéndice E, Ficha Técnica de Briquetadora. Piura: Institucional repository PIRHUA; 2014

[9] Orrego Moya R. En Estado del arte y novedades de la Bioenergía en el Perú. La Bioenergía en América Latina y el Caribe. Santiago: Organización de las Naciones Unidas para la Alimentación y Agricultura (FAO); 2013

[10] Espinoza C, Bernabel F. Emisiones al ambiente generadas por la transformación de energía, consumo propio y consumo final de energía comercial. Balance Nacional de Energía (BNE). Lima: Ministerio de Energía y Minas del Perú (MEM); 2014

[11] Velásquez J. Evaluación del Potencial de Generación Energética con cáscara de arroz en la zona del Huallaga Central del Departamento de San Martín. Lima: E-Publishing of the Regional Direction of Energy and Mines; 2013

[12] Briquetas Combustibles para uso doméstico. In: Norma Técnica Colombiana; 2013

[13] Fonseca E, Tierra L. Desarrollo de un Proceso Tecnológico para la obtención de Briquetas de aserrín de madera y cascarilla de arroz, y pruebas de producción de gas pobre. Riobamba: Escuela Superior Politécnica de Chimborazo (ESPOCH); 2011

[14] Tolosana E, Ambrosio Y, Laina R, Martínez R, Pinillos F. Rendimientos y Costes de diferentes aprovechamientos de la biomasa forestal. Madrid: Institute for the Diversification and Saving of the Energy and Mines; 2007

[15] Botta NA. Salud e Higiene laboral. In: Poder Calorífico. Buenos Aires: Red Proteger; 2012

[16] Velásquez Piñas J. Evaluación del potencial de generación energética con cáscara de arroz en la zona del Huallaga Central del departamento de San Martín. Moyobamba: Editorial of the Regional Direction of Energy and Mines San Martín; 2014

[17] Orts B, Campos B. Métodos de análisis granulométrico. Aplicación al control de la granulometría de materias primas. Castellón: E-Publishing of the Association of Research of the Ceramic

Industries, Instituto de Tecnología Certimica Universitat Jaume 1; 2008

[18] Faborode M, O'Callaghan J. Optimizing the compression briquetting of fibrous agricultural materials. Journal of Agricultural Engineering Research. 1987:245-262

[19] Bailis R, Ogle D, MacCarty N, Still D. The Water Boiling Test (WBT). Household Energy and Health Programme; 2007

[20] Lubwama M, Yiga V. Characteristics of briquettes developed from rice and coffee husks for domestic cooking applications in Uganda. Renewable Energy. 2017. DOI: 10.1016/j.renene.2017.11.003

[21] De Olivera R, Palacio S, Da Silva E, Mariani F, Reinehr T. Briquettes production for use as power source for combustion using charcoal thin waste and sanitary sewage sludg. Environmental Science and Pollution Research. 2017. DOI: 10.1007/s11356-017-8695-0

[22] Sarangi M, Bhattacharyya S, Behera RC. Rice Effect of Temperature on Morphology and Phase Transformations of Nanocrystalline Silica Obtained from Rice Husk2012. pp. 377-386

[23] Giwa A, Alabi A, Yusuf A, Olukan T. A comprehensive review on biomass and solar energy for sustainable energy generation in Nigeria. Renewable and Sustainable Energy Reviews. 2017;**69**:620-641

[24] Assureira E. Rice husk, an alternative fuel in Peru [Abstract]. In: Biomass and Coal Research Programme. Lima: E-Publishing of the Pontifical Catholic University of Peru; 2002

[25] Obi OF, Ezema JC, Okonkwo WI. Energy Performance of biomass cookstoves using fuel briquettes. In: AIP Conference Proceeding. 2017. DOI: 10.1063/1.5021237

[26] Brand M, Jacinto R, Antunes R, Bayestorff A. Production of briquettes as a tool to optimize the use of waste from rice cultivation and industrial processing. Renewable Energy. 2017. DOI: 10.1016/j.renene.2017.03.084

[27] Lozada Cubas J. Evaluación de la reducción de uso de leña, para la producción de ladrillo arcilla. Tarapoto: E-Publishing of National University of San Martin; 2012

[28] La Tienda Biomasa. Revista Leñas Oliver. 2015. Extracted on: http://tiendabiomasa.com/medir-lena

[29] Estudio Técnico de la Leña de la Sede Forestal Moyobamba. Moyobamba: Editorial of Regional Government San Martín; 2009

[30] Torres H, Agreda J, Polo C. Evaluación de impacto ambiental producido por el uso de cocinas tradicionales en el área de conservación regional Vilacota-Maure, Tacna, Perú. 2012. DOI: 10.23850/22565035.25

[31] Prada A, Cortés C. La descomposición térmica de la cascarilla de arroz: Una alternativa de aprovechamiento Integral. Revista Orinoquia. 2010;**14**:155-156

Chapter 7

# Voluntary Certification of Carbon Emission in Brazil - The Experience of an Electricity Trader

*Fernando Amaral de Almeida Prado and Edvaldo Avila*

## Abstract

Few countries in the world have such availability of natural resources as Brazil. Even so, the country records increasing greenhouse gas (GHG) emissions related to electricity, and this is due to political and economic factors. This chapter shows the experience of the largest Brazilian power trader in its pioneering effort to develop voluntary certifications (2011) in power buy and sell transactions, along with other energy efficiency actions. The initiative has accumulated 9 years' experience with more than 1600 units in different industries, using a methodology aligned with the Paris Agreement. The chapter presents the calculation methodology and the safeguards that ensure information integrity and verification of the certified indicators. Only renewable sources are used in this methodology, such sources being qualified as incentivized by their sustainability characteristics being small-size power plants (less than 30 MW of capacity installed).

**Keywords:** greenhouse gases, voluntary certification, power trading, Paris Agreement Brazil

## 1. Introduction

The creation of a project for voluntary certification associated with the consumption of electrical energy developed jointly by Sinerconsult Consultants and Comerc Energia, the largest power trading company in Brazil (Comerc manages a portfolio around 26,000 GWh/year), was motivated by the perception of the worsening emission conditions related to the Brazilian energy sector. Both organizations were among the first to realize in Brazil that Kyoto Protocol Policies would lose force, and therefore, the natural alternative would be to adopt voluntary measures, as the Paris Agreement would later prove during Conference of Parts (COP 21).

Few countries in the world have such a strong renewable energy generation matrix as Brazil. **Figure 1** shows that 73% of the installed capacity in Brazil comes from renewable sources [1]. In the past 5 years, however, power generation from thermal plants has been on the rise, despite the growing proportion of renewable energies. **Figure 2** indicates this growth [2].

This can be traced back to a conceptual issue regarding environmental protection. Since the mid-1990s, all new hydroelectrical power plants have been conceived as run-of-river.

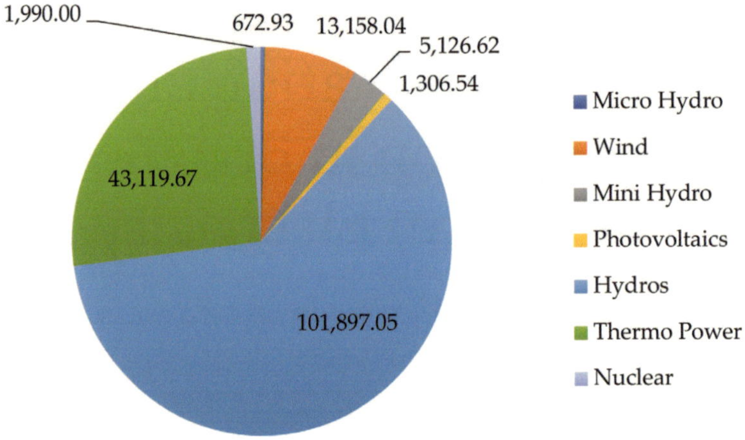


**Figure 1.**
*Capacity power in Brazil in MW (July 2018) [1].*

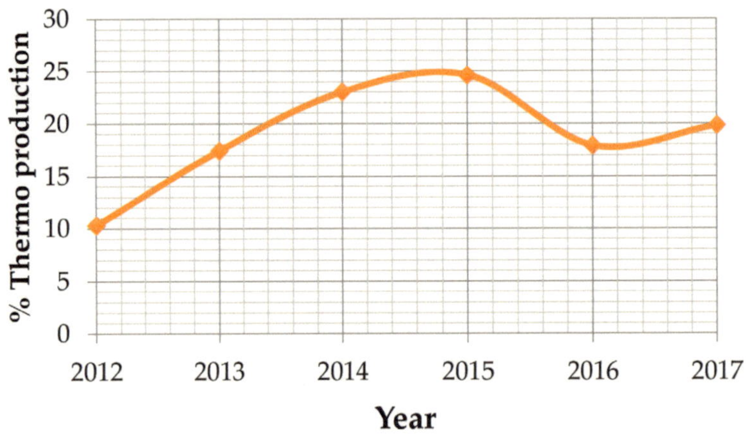


**Figure 2.**
*Production percentage from thermal power plants [2].*

Therefore, the operational capacity of the Brazilian hydroelectrical power plants, once fed by large rivers, no longer had pluriannual reservoirs. More and more, whenever hydrological conditions are adverse, the interconnected system needs to rely on thermal plants. Another point is that the growing insertion of wind farms, characterized by high intermittence, also leads to more frequent deployment of thermal plants. As a result, greenhouse gas emissions (GHG) have grown significantly, albeit still relatively low compared to countries with high thermal generation profiles. Alarmingly for Brazil, though, the emission levels measured in 2014 are already higher than government projections for 2030 [3]. **Figure 3** shows the growth trend for emissions [4].

Paradoxically, one of the countries with the broadest natural resources has shown deteriorating performance in emission indicators correlated with climate change.

As mentioned above, the perception of the increasing importance of voluntary actions and growing emissions attributable to the electrical energy industry have been the two key driving forces that led to the actions that will be detailed in the following sections. Nevertheless, from a wider perspective, the key issue is climate change, a serious concern for most countries.

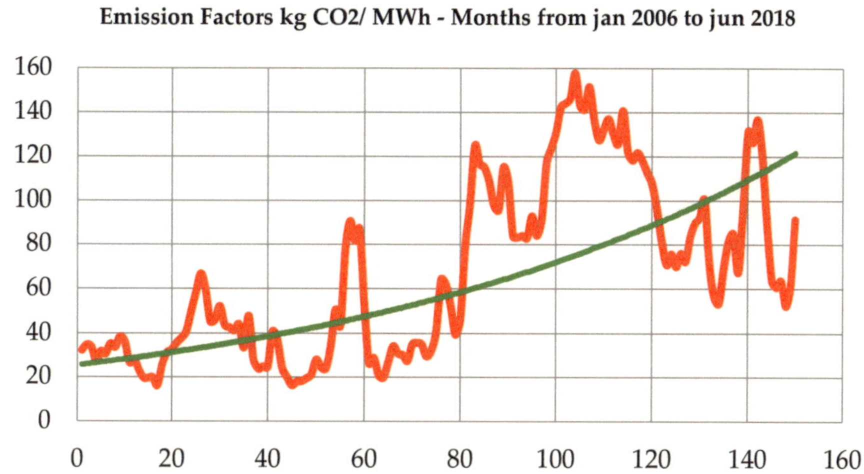


**Figure 3.**
*Emission factors for the interconnected Brazilian system [4].*

Section 2 presents the analysis of voluntary markets and the need for clearly defined certification. Section 3 reviews the general context of the Paris Agreement and its pending issues regarding emission certification. Section 4 analyzes the general problem of calculating the greenhouse gas emissions in interconnected systems. Section 5 presents the factors that led Sinerconsult and Comerc to develop the certificates in the proposed form and the concepts utilized. Section 6 describes the calculation methodology. Section 7 reports the results achieved during the 9 years since the implementation of the certificate. Section 8 describes the evolution from certificate to energy efficiency actions, and Section 9 presents the final tentative ideas for the future and perspectives for voluntary certification in Brazil.

Several ideas related to this chapter was developed initially in an article "Clean Energy Certification in Brazil: A proposal," published in the Journal of Sustainable Development of Energy, Water and Environmental Systems, by two of the authors[1] of this chapter [5].

## 2. Voluntary markets: Do they work?

Over the years, many companies have received incentives or mandatory rules in order to develop initiatives to reduce their GHG emissions.

A voluntary market is one that comes from a no mandatory initiative, decided by a country or by a corporation in order to make a sensitive contribution to one "cause."

In this chapter, the focus is centered on an initiative that contributes to reduce GHG emissions and to help in the fight against climate change made by volunteered initiative of one of the biggest energy trader in Brazil.

Even the USA, a country that has not to date adhered to formal agreements and remains adamant in resisting international cooperation, has hundreds of voluntary initiatives of their own to reduce emissions. Here, we list some of the strategies that have been used: (i) regional legislation, (ii) sectorial policies, (iii) initiatives by industry associations, unions, and nongovernmental organization (NGOs), and (iv) business initiatives. Each one of the initiatives has its own motivation, and they may or may not be limited to their corresponding industry.

[1]The other two authors of original paper allowed the present authors to use the primary information.

Take for example, the carbon disclosure project (CDP) [6], which in 2017 involved more than 6300 of the largest companies around the world to reveal progress in avoided emissions. Around 89% of them now have their own emission reduction target. Such initiatives can foster best practices around the world by helping people and companies think strategically about climate change. More than this, most of the companies also included their suppliers in their reduction targets. If met, the targets could be relevant contributions to the required GHG abatement to cap global warming at no more than +2°C. The CDP report disclosed reductions equivalent to 551 million tons of $CO_2$ in 2017, with associated cost savings of US$ 14 billion [6].

A review of the literature on voluntary markets [7–16] indicates that when information is available, the behavior of customers could be affected, also the demand for environmental friendly products. Companies with voluntary initiatives can benefit from them by gaining a positive image with final consumers.

Delmas and others [10] found that once energy is required to produce goods and services, consumers can drive change by choosing goods and services associated with renewable or "green" energy [10]; a good example of its importance is that renewable energy market in United States of America (USA) in the last 10 years (2007–2017) has grown up 5.4% yearly [16].

Delmas and others [9] found that a deregulated industry where competition is still incipient will be more affected by consumer perception and sensitivity to the issue, favoring the insertion of renewable sources, as can be better explained in following sections. The same perception can be reported by the authors of this chapter, in their experience in Brazil. Some of the clients of Comerc are very proud about their certificates. However, Delmas and others [9] note that sources of low cost such as coal can affect the decision process with the low price being the winner.

Kotchen, on the other hand [11], put on doubt if simple low-cost public policies can effectively promote stable voluntary initiatives, and whether such initiatives will continue to be effective, especially if more centralized policies are required in future.

Are voluntary and mandatory initiatives complementary or substitutes? In the opinion of authors, and likely in the opinion of anyone who reads the 2018 CDP Report [6], all parties, government and customers, must be involved in the effort. The figures cited by Hamilton [14] (volunteer markets could be US$ 100 million/year) probably will be much more impressive with a successful Paris Agreement.

The initiatives give companies the tools they need to be prepared to lead the way in GHG regulation. "This market is growing fast, perhaps doubling each year." Hamilton's recommendation, however [14], centered in the needs of tools to measure the emissions targeted, an opinion that the authors share. This need comes from a pattern that will be required for compliance of the goals self-established.

There is no unanimity about the efficiency of volunteer markets; Ferguson, by instance [15], believes in many barriers, high costs, and complexity in reporting trustable results. Higher costs in voluntary markets result from the nonexistence of deterministic goals, as obvious. The agent can always decide not to invest in reducing its emissions, while others in doing so make its operation more costly. In regulated industry, usually the regulator does not recognized costs that are not strictly related with regulation, so even an action that could be defensible may impact in economic results of a goodwill initiative.

Reporting results is always complex. There are several alternatives in the way to report the figures, per unit of production or through corporations and their subsidiaries, especially in different countries with different legislations. Some emissions come directly from the company and the others from the suppliers in chain production. Due to avoiding double accounting, this kind of information must be carefully reported and checked.

More than this, there are the difficulties related to the "leaks" that can occur along the chain production. Just to focus on the electricity industry, one can use the example of a wind power plant (carbon-free by definition) but that needs a high voltage line of interconnection, which demands some deforestation.

Kim and Lyon [12] share the pessimist view because only projects with low marginal costs are likely to succeed, as the regulatory risk is too high. Regulatory risks should be especially considered because if there are no rules, a subsequent emergence of these rules can make impossible to account for past initiatives. The hypothesis of the emergence of rules is not contradictory even in voluntary markets, since a country can voluntarily create its goals in international diplomacy, and to accomplish them, needs to encourage their fulfillment by incentive or even by mandatory regulation in some segments of the economy. A cap and trade environment could help the management of this kind of uncertainty.

Other beliefs from Hofmann [8] are connected with the junction of a public policy with some associated benefit, such as Brazilian incentives in transmission tariffs for small renewable generators, and volunteer markets. The authors would add to this, the goodwill or favorable image associated with environmental marketing.

## 3. Paris Agreement: understanding the main issues

Two decades after the creation of the United Nations Framework Climate Change Convention (UNFCCC), the parties (countries) remain firm in their decision to contribute to the reduction of GHGs, but the debate continues about how to share the burden among the parties, especially because of the substantively different development levels. The problem of coordinating actions has also been considered very important.

The previous experience with the Kyoto Protocol, where responsibility was distributed differently between developed and undeveloped countries, did not work. Countries such as the United States and Canada did not ratify the protocol in the belief that the effort and cost faced by developed countries would be wasted by developing countries, which had no deterministic emission reduction targets at the time [17].

The established targets and metrics were also questioned, so after the Kyoto Protocol expired, no new agreements with similar methodologies could be established. An interesting example of such divergences may be the case of China: the country's emission indicators are higher than those in the USA, yet are substantially smaller if considered on a per capita basis. A parallel reasoning can be applied to the analysis if we consider the useful life of emissions in the atmosphere: in a cumulative calculation, Chinese emissions remain far below those produced over decades by developed countries.

This deadlock was bypassed in the Paris Agreement, which declares the sovereignty of each country in choosing and setting possible goals within a process of goodwill. Article 6 of the Paris Agreement states that countries may cooperate internationally in different ways in order to reach climate goals and defines broad enough conditions, so that the targets set voluntarily by the parties, known as Nationally Determined Contributions (NDC), can be achieved.

The UNFCCC targets established that the average world temperature not exceed 2°C at least with a probability of 50%, so the goals for different countries, even defined in voluntary way, need to take in count a common objective.

There is a mood of relative optimism, or at least, it seems to be overcoming the pessimism that followed the expiration of the Kyoto Protocol; however, there are still many adjustments to be made.

Among the most relevant points that could be mentioned is the need to create metrics to compare the efforts expended by countries, since different ethical concepts can be raised, involving, for example, the "polluter pays" principles [18] (especially defended by Brazilian Diplomacy).

Other aspects include the principle of equity, in other words, the right each party has to guarantee its citizens can have access to the planet's natural resources; the principle of capability, the capacity to produce actions that are feasible for the country, and finally, the principle of sovereignty, that involves the discussion of whether countries should have proportional targets or whether the sovereign right to decide according to their circumstances would apply. It should be noted that with regards to the "polluter pays" principle, it would be necessary for the carbon "price" to be evenly defined to avoid "polluter havens" [16].

Most of the points above are included in the COP 24 agenda—the summit will take place in Poland in 2018 to discuss accounting principles, legislation, procedures, compliance with the targets defined by the NDCs themselves, as well as the rules to report reductions achieved through market mechanisms, including voluntary certifications.

## 4. Grid emissions: understanding the problem

Brazil has one of the largest interconnected systems in the world with similar dimensions to Western Europe (**Figure 4**); for large systems like this, it is very hard to quantify the amount of GHG emitted, especially with a combination of so many different sources with different environmental attributes like a coal-burning thermal plant or a carbon neutral small-scale wind turbine, for instance.

**Figure 4.**
*Brazilian interconnected grid [4].*

Being the Brazilian commercial model, a system of free competition, theoretically each plant at any point of the grid could inject the energy destined for a certain final consumer. It happens that even though the system is interconnected, the laws of physics determine that power and electric flows occur depending on the network topology, voltage levels, and the relative positioning between generation and load. Also, as interconnected grids are operated usually by independent entities and the criteria for dispatch is in regular bases, efficiency and supply security, it is virtually impossible to unequivocally associate load and generation.

This could be more complicated in Brazil, as sometimes hydro plants (the main source of power production in Brazil) may even be switched off to preserve water in the drier months, for future use. The resulting energy deficit is offset by energy produced by thermal plants burning a range of fossil fuels.

Therefore, there would be no one to one correspondence between generation and consumption, so the emission factor likely to be accounted for could be only that resulting from the average value to be determinate from some reasonable criterion.

However, there is a very important conditioner that brings a solution to this issue, the purchase and sale contracts. Thus, considering that, the energy produced cannot be stored (at least not in relevant quantities and at competitive costs) and assuming that the electricity consumed is equal to that produced (after technical and commercial losses), the match between generation and load is supported by the contracts. Therefore, the generating fact that connects a consumer to a low-emission production (for example, small hydro power plant) can be made through the contract between the parties.

The methodology used by Comerc-Sinerconsult uses the set of rules established by the United Nations (UN) named "ACM-002-Methodology for Calculating the Average Grid Emission used for Clean Development Mechanism," available on the UNFCCC website [19]. Even if considering that the Kyoto Protocol is no longer valid, the methodology, which was developed with sound principles, has criteria that remain valid. By the way, it is based on this methodology that the Brazilian Government through its Ministry of Science Technology Innovation and Communication (Brazilian Designated National Authority—DNA) publishes monthly the hourly statistics of the emissions of the electric grid. These statistics are published for both carbon credit projects, whose useful life still remains after the Kyoto Protocol and for corporate inventories [20].

The methodology discussed in this chapter is the one that is destined to inventories which reflect the Brazilian emissions on time line. It is noteworthy that unlike other countries, the Brazilian emissions, which are naturally very low among countries around world, have been worsened due to the massive insertion of intermittent renewable sources in the Brazilian electricity matrix (they need more thermo power plants in backup reserves) and by the growing difficulties of the hydrological regimes in the last 4 years.

Since 2009, Comerc using the methodology developed by itself and in partnership with Sinerconsult served more than 1600 electricity consumers with power from incentivized sources in the deregulated market. This portfolio of clients has companies of more different industries in Brazil, like chemicals, vehicles, and auto parts, food, surgical and hospitals, electroelectronics, household cleaning products, packing, personal care, paper and cellulose, leverage, and so many others.

Brazil decides that hydro plants of any size, biomass thermal plants, solar farms, wind facilities, and some qualified cogeneration plants must be considered as carbon neutral. Although it is a known fact that some hydro plants do emit greenhouse gases, the Brazilian DNA has determined that for accounting purposes in the Brazilian electric power sector, all hydro plants are to be considered as having no GHG emissions.

## 5. Drives to voluntary certification: Comerc/Sinerconsult

In 2008, while the severity of the climate change became increasingly clear, no companies or customers in Brazil seemed aware of the issue, apparently believing that a country with plentiful natural resources has no reason for concern. At the time, they underestimated the fact that the strong presence of hydro power plants was built in the 1960s and 1970s, and could not be considered under the Kyoto Protocol and its eventual outcomes. Ironically, if electric power plants of any age could be considered, Brazil could expect that Amazon Forest will be listed as a contribution to efforts to combat climate change.

They also disregarded the fact that the new run-of-river hydroelectrical plants developed since the 1990s would increasingly require backup from thermal plants to ensure safety and to meet operational requirements during years of unfavorable rainfall. The same mistake was made when evaluating the intermittence of wind and solar power plants (solar plants still incipient at that time).

The idea of volunteer certification was inspired by the Conference of Parties (COP), which strived for consensual decisions—a daunting challenge given the diversity of political regimens and the cultural structures of each party. A tongue-in-cheek remark—we all know how hard it is to reach consensus when allocating parking spaces in a condo homeowners' meeting, so one can only wonder about the chances when discussions involve such disparate parties. Time showed that volunteering was the winning idea in COP 21.

In a pioneering initiative in Brazil, Comerc and Sinerconsult launched certificates for avoided emissions based on the consumption of renewable energy (incentivized energy as it is called in Brazil). Our priority at the time was to create a process that was easy enough to be understood by the players, while robust enough and equipped with safeguards to ensure ethics, reliability, transparency, traceability, and coherence. All the information used is based on official data from energy contracts and their validation, as explained in the next section.

The measurement presented in the certificates, that is, tons of $CO_2$ avoided, might not be easily understood among lay audiences, so an indicator of equivalent reforestation was added, making it easier for the general audience to understand the dimensions of the avoided emissions by comparing it to a certain number of trees planted.

Obviously, reforestation figures could be very different numbers depending on tree types, harvest period, and spacing. For clarity, the certificate established a standard reforestation model, using calculations for avoided emissions approved for a project in the Clean Development Mechanism (CDM) in Brazil. This way, every calculation for equivalent number of trees follows the same conceptual basis, coherent with the Unite Nation International Panel of Climate Change (IPCC).

Although this methodology was created many years before the Paris Agreement, the conceptual directives defined do not conflict with the new adopted principles. Finally, it is necessary to point out that the GHG Protocol, one of the most important references in certifications around world, known to adopt conservative positions in the linkage of power plants and consuming units, adopted from 2017 a similar assumption as the Comerc-Sinerconsult model. The contract is the originating fact in establishing the environmental quality of the energy provided, to a consumer.

## 6. Methodology for modeling voluntary-certified avoided emission

By law, in Brazil, small power plants using renewable resources and with low environmental impact have a financial incentive in the form of discounted energy transport rates (TUSD—Distribution System Usage Tariffs, in the Portuguese acronym). In other words, they pay a lower tariff for using the grid systems when the energy that was consumed is provided by an incentivized source.

The legislation establishes that the Regulatory Agency in Electricity Industry (ANEEL—National Electric Energy Agency, in the Portuguese acronym) must stipulate a tariff reduction of no less than 50% for transport of energy that comes from small hydro power plants, photovoltaic farms, wind power plants, and biomass-fueled thermal plants (in special sugarcane bagasse), as well some qualified cogeneration, all of plants with capacity smaller than 30 MW.

It must be detached that the benefits are allowed also for final customers. The generation facilities pay 50% of transport tariff from their site until the gravity center of the electric system and the customers pay 50% from gravity center to its location. This is in line with the ideas of Hoffman [7] discussed in Section 2.

All eligible plants for these rebates are environmentally friendly and are considered carbon neutral. Consequently, identifying a plant that has discounts is similar to identifying a source of zero emissions. Here, it is possible to identify a two-way match. The problem remains as how to ensure that the energy actually comes from a set of incentivized power plants.

To solve this issue, the information provided by the Electric Energy Trading Chamber (CCEE, in Portuguese acronym) is fundamental. The CCEE is the organization responsible (officially) for the supervision and control of electric energy trades among distributors, traders, free consumers, and generators in Brazilian market. In short, CCEE is a clearing house for electricity contracts in Brazil.

Transactions based on incentivized energy are eligible for discounted transport tariffs, so the subsidize is allocated to tariffs of all other consumers that do not use renewable of small plants [5]. For this reason, it is very important the perfect identification of whom is eligible for the discounts, because the bigger they are, the more they impact other consumers. The regulator is quite worried about the fiscalization of subsides.

The Regulation Agency (ANEEL) established that CCEE is the entity responsible to assure that the energy traded with discounts comes from a source eligible by law for this kind on incentives. Since January 2009, the CCEE has consistently published an index known as the "discount matrix," with the correlation between consumers and incentivized energy.

As related by de Almeida Prado et al. [5], the information provided from CCEE uses criterion of governance that gives confidence to stakeholders about the "quality" of energy used in each unit of consumption. The information could be checked by anyone to ensure about its reliability. All the figures are traceable and auditable and the rules are stable in time line. If a block of energy is tradable from an incentivized source and deserves the discount, we can assure that this amount of energy comes from a GHG neutral source [5].

Thus, this methodology indirectly uses an official source to determine what percentage of the power consumed by a specific consumer actually comes from a GHG neutral source. One should remember that there is always the possibility that a given power plant will be unable to produce all of the energy sold. In such situations, the generator or trader must purchase energy from third parties to honor its agreements and provide the energy it sold but is unable to deliver. If this "replacement" energy comes from other sources such as a nonincentivized, the consumer loses the right to the discount, in same proportion to the "not green" amount of energy supplied. The loss of this discount is made official by the CCEE and this procedure avoids that incorrect subsidizes could damage other stakeholders [5].

The proposed methodology uses an indirect but official inspection tool, which identifies the proportionality of the energy with incentives and therefore from sources that have zero emissions or are GHG neutral. This methodology determines how much of the energy consumed is eligible for a transport discount and reduces the emissions published by the government for that particular month by a proportional amount.

The outcome is supported by the reliability of two official sources, one the amount of GHG emitted by the grid, and another by the exact volume of electricity consumed that was generated from renewable, GHG neutral sources. This reliability extends to the period during which the data are calculated, as both indicators are calculated for each calendar month, avoiding any distortions related to the period of calculation of these indicators.

The method used to calculate these numbers is described below. It is based on the trading chamber (CCEE) "ME001" (energy consumed) and "EI002" (TUSD incentive discount) reports.

First, the weekly consumption of energy reported in ME001 (energy consumed) reports is added up to come up with the total for the month. The amount of energy traded at a given percent discount is added and divided by the total volume, to arrive at:

$$TD = \frac{\Sigma VE * D}{\Sigma VE} \tag{1}$$

where *TD* is the total discount, *VE* is the volume of energy, and *D* is the discount.

Consumption is then multiplied by the discount to arrive at the incentive that applies to the volume of energy:

$$\text{MIAE} = \Sigma \text{WE} * \text{TD} \tag{2}$$

where *MIAE* is the monthly incentive applicable energy and *WE* is the weekly consumption.

The difference between monthly consumption and the amount of incentivized energy is then used to calculate the GHG emissions avoided each month, reported as tons of $CO_2$ equivalents. This is calculated as a specific emission factor such as tons of $CO_2$eq/MWh:

$$\text{AE} = (\text{TMC} - \text{IAEC}) * \text{EF} \tag{3}$$

where *AE* is the avoided emissions, *TMC* is the total monthly consumption, *IAEC* is the incentive applicable energy consumption, and *EF* is the emission factor.

The procedures described herein abide by the generally accepted principles for calculating inventory, which are relevance, universality, precision, transparency, and consistency. All of them are connected with the good practices presently discussed in Paris Agreement. Calculating avoided GHG emissions is a simple and reliable process if one has access to the customer reports issued by the CCEE regarding electricity consumption, specifically ME001 and EI002. Such reports are available only for customers, but of course they can, if necessary, give open access to anyone charged with checking the figures [5].

## 7. Results

The results obtained are substantial and represent a pioneering initiative in voluntary measures to reduce GHG emissions in Brazil. **Table 1** presents the results of 900 different companies, with more than 1600 consumer units that have been using this methodology since 2009.

| Year | Number of certificates | Ton $CO_2$eq | Number of equivalent trees |
|---|---|---|---|
| 2009 | 75 | 21,279.70 | 121,787.44 |
| 2010 | 75 | 76,900.86 | 440,117.14 |
| 2011 | 120 | 66,334.57 | 358,203.24 |
| 2012 | 192 | 111,248.36 | 778,738.53 |
| 2013 | 385 | 344,337.79 | 2,410,364.53 |
| 2014 | 474 | 528,496.53 | 3,699,475.71 |
| 2015 | 326 | 550,516.87 | 3,853,618.09 |
| 2016 | 996 | 473,668.40 | 3,315,678.82 |
| 2017 | 1130 | 701,854.64 | 4,912,982.48 |
| Σ | 3773 | 2,874,637.72 | 19,890,966 |

**Table 1.**
*Figures of Comerc Sinerconsult certificates (2009–2017).*

## 8. Next steps

As discussed above, climate change is increasingly becoming a serious issue in light of the severe effects it might produce in human life. The pioneering Comerc-Sinerconsult initiative is not the only option available today. There are other initiatives, for example, from the Brazilian Society for Wind Power (Abeeolica, in the Portuguese acronym), and from associations of sugar and alcohol producers, and even international entities, for example, GHG Protocol and the International REC Standard.

It is clear that ongoing regulatory follow-up must be part of all joint activities undertaken by certificate sponsors, who should be open to include enhancements and committed to the continuous improvement of the project.

Since 2017, Comerc developed similar concepts for the certification of energy efficiency. The emission factors in this case are obviously not the same as those utilized in corporate inventories neither the conserved energy is defined by the Chamber of Energy Commercialization (CCEE). However, the methodologies developed were maintained regarding ethics, reliability, transparency, and coherence. Comerc also maintained the equivalence with reforestation for a clearer presentation of figures to a nonexpert audience. The first certificates were already checked and will be expanded quickly as the energy efficiency actions are more valued in Brazil.

As the concepts from the Paris Agreement become consolidated, it will be possible to develop new activities with the "potential" commercialization of certificates and its utilization for the neutralization of events or transferences among companies of the same group. Any such steps will be developed with the caution that characterized the creation of the certificates.

## 9. Conclusion

In the authors' opinions, voluntary certification represents a path of no return for public projects and policies related to climate change.

The references presented in the Section 2 section indicate that consumers may exert pressure on the supply chain in different markets. This perception is aligned with Comerc experience in Commercial Relations and Marketing: more than 900 corporations receive the emission certificates today.

No one of these initiatives are easy to control, the visited literature indicates the need to take care of the metrics calculation, so that it is possible to offer reliability to the stakeholders in the use the data of these voluntary initiatives, for commercial planning, company records, environmental compliance reports, or commercial and marketing policy actions.

The methodology described by the authors brings in their control, mechanisms very robust and criteria that offer security and reliability in the figures obtained.

The methodology proposed by Comerc-Sinerconsult was the pioneer in Brazil for this type of action. Given the theme's importance, innumerous other initiatives have arisen since the first certificates were emitted in 2009. It is important to note that the GHG Protocol, important reference that internationally had very strict criteria for the accounting of GHG emissions in interconnected grids began to use criteria similar to the Comerc-Sinerconsult since 2017. It demonstrates how important voluntary initiatives are, because they promote learning by the need to create and develop pioneering criteria and end up transforming the market in an evolutionary sense. The very transformation of a set of mandatory rules originating in Kyoto seems to find more appropriate conditions for its success now with the voluntarism of the Paris Agreement.

This chapter presented the pioneering initiative led by Comerc and Sinerconsult, creating the first avoided emission certificates in Brazil. Since 2009, we have reported almost 3 million tons of equivalent $CO_2$ that are no longer released into the atmosphere due to the commercialization of renewable energy from small-scale projects with low or no environmental impacts. This volume of emissions is equivalent to a reforestation of approximately 20 million of trees and involved the participation of 900 companies with more than 1600 consumer units, all of them Comerc clients purchasing renewable energy or undertaking energy efficiency actions.

New perspectives are open now with the Paris Agreement that priories volunteer initiatives. The authors believe that in short time new markets of certification will result from similar initiatives. The path probably will be the commercialization of certificates and its utilization for the neutralization of events or transferences among companies of the same group. Any such steps must be developed with the caution that characterized the creation of the certificates here described.

## Author details

Fernando Amaral de Almeida Prado[1]* and Edvaldo Avila[2]

1 Sinerconsult Consultoria Treinamento e Participações Limitada, São Paulo, Brazil

2 Comerc Energia, São Paulo, Brazil

*Address all correspondence to: fernando@sinerconsult.com.br

## References

[1] ANEEL—Agência Nacional de Energia Elétrica. 2018. Available from: http://www2.aneel.gov.br/aplicacoes/capacidadebrasil/capacidadebrasil.cfm [Accessed: 24-07-2018]

[2] ONS—Operador Nacional do Sistema Elétrico. Available from: http://ons.org.br/paginas [Accessed: 24-07-2018]

[3] Cursino A. Emissões de $CO_2$ pela geração de eletricidade no Brasil superam em 2014 a previsão do governo para o ano de 2030. Available from: http://mitsidi.com/emissoes-de-co2-pela-geracao-de-eletricidade-no-brasil-superam-em-2014-a-previsao-da-epe-para-o-ano-de-2030/?lang=pt-br [Accessed: 24-07-2018]

[4] MCTIC—Ministério da Ciência Tecnologia Inovação e Comunicação. Fatores de Emissão para inventários Corporativos. Available from: http://www.mctic.gov.br/mctic/opencms/ciencia/SEPED/clima/textogeral/emissao_corporativos.html [Accessed: 24-07-2018]

[5] de Almeida Prado FA Jr, Rodrigues da Silva AL, Avila EM, Matsuyama G. Clean energy certification in Brazil: A proposal. Journal of Sustainable Development of Energy, Water and Environment Systems. 2015;**3**(1):95-105. DOI: 10.13044/j.sdewes.2015.03.0007

[6] CDP—Carbon Disclosure Project. Available from: http://b8f65cb373b1b7b15feb-c70d8ead6ced550b4d987d7c03fcdd1d.r81.cf3.rackcdn.com/comfy/cms/files/files/000/001/761/original/CDP_IMPACT_INFOGRAPHIC-FINAL-2.jpg [Accessed: 24-07-2018]

[7] Desgagné BS, Gozlan E. A theory of environmental risk disclosure. Journal of Environmental Economics and Management. 2003;**41**:377-393. DOI: 10.1016/S0095-0696(02)00056-6

[8] Hoffmann A. Climate Change Strategy: The business logic behind voluntary greenhouse gas reductions. Working Paper Ross School of Business; 2004

[9] Delmas M, Russo MV, Montes-Sancho M. Deregulation and environmental differentiation in the electric utility industry. Strategic Management Journal. 2007;**28**:189-209. DOI: 10.1002/smj.578

[10] Delmas M, Montes-Sancho M, Shimshack J. Information disclosure policies: Evidence from the electricity industry. Economic Inquiry. 2010;**48**(2):483-498. DOI: 10.1111/j.1465-7295.2009.00227.x

[11] Kotchen M. Climate policy and voluntary initiatives: An evaluation of the Connecticut Clean Energy Communities Programs. Working Paper 16.117. National Bureau of Economic Research; 2010

[12] Kim EH, Lyon TP. Strategic environmental disclosure: Evidence from DOE's voluntary green house gas registry. Journal of Environmental Economics and Management. 2011;**61**:311-326. DOI: 10.1016/j.jeem.2010.11.001

[13] Bisore S, Hecq W. Regulated (CDM) and voluntary carbon offset schemes as carbon offset markets: Competition or complementarity? Working Paper Centre Emile Bernheim, Solvay Brussels School of Economics and Management; 2012

[14] Hamilton K, Schuchard R, Stewart E, Waage S. Offsetting emissions: A business brief on the voluntary market. Ecosystem Marketplace. Second edition, (2008). Available from: http://www.bsr.org/reports/BSR_Voluntary-Carbon-Offsets-2.pdf [Accessed: 12-5-2013]

[15] Fergurson J, Harris J, Hart JS, Ramakrishnan K, Thompson T, Weber S. Voluntary Greenhouse Gas Reduction Programs have Limited Potential. Report 08-P-0206, US Environmental Agency, Report 08-P-0206. 2008

[16] Renewable Energy. Available from: https://www.eia.gov/totalenergy/data/monthly/pdf/sec10.pdf [Accessed: 20-08-2018]

[17] Sheriff G. Burden sharing under the Paris Agreement. Working paper#16-04. National Center for Environmental Economics; 2016

[18] Clark D. What is the "polluter pays" principle. Available from: https://www.theguardian.com/environment/2012/jul/02/polluter-pays-climate-change [Accessed: 25-7-2018]

[19] UNFCCC—United Nations Framework of Climate Change Convention. ACM-002. Available from: http://cdm.unfccc.int/methodologies/DB/M0CSBFOF8RQG5I84XU5Y4WX0I5LHS1 [Accessed: 12-5-2013]

[20] MCTIC—Ministério Ciência Tecnologia Inovação e Comunicação. Available from: http://www.mctic.gov.br/mctic/opencms/ciencia/SEPED/clima/textogeral/emissao_corporativos.html [Accessed: 25-07-2018]